FREE Test Taking Tips DVD Offer

To help us better serve you, we have developed a Test Taking Tips DVD that we would like to give you for FREE. **This DVD covers world-class test taking tips that you can use to be even more successful when you are taking your test.**

All that we ask is that you email us your feedback about your study guide. Please let us know what you thought about it – whether that is good, bad or indifferent.

To get your **FREE Test Taking Tips DVD**, email freedvd@studyguideteam.com with "FREE DVD" in the subject line and the following information in the body of the email:

a. The title of your study guide.

b. Your product rating on a scale of 1-5, with 5 being the highest rating.

c. Your feedback about the study guide. What did you think of it?

d. Your full name and shipping address to send your free DVD.

If you have any questions or concerns, please don't hesitate to contact us at freedvd@studyguideteam.com.

Thanks again!

SSAT Upper Level Prep Books 2020 & 2021

Upper Level SSAT Prep & Practice Test Questions for the Secondary School Admission Test [5th Edition]

Test Prep Books

Written and edited by Test Prep Books.

ISBN 13: 9781628457827
ISBN 10: 1628457821

Table of Contents

Quick Overview

As you draw closer to taking your exam, effective preparation becomes more and more important. Thankfully, you have this study guide to help you get ready. Use this guide to help keep your studying on track and refer to it often.

This study guide contains several key sections that will help you be successful on your exam. The guide contains tips for what you should do the night before and the day of the test. Also included are test-taking tips. Knowing the right information is not always enough. Many well-prepared test takers struggle with exams. These tips will help equip you to accurately read, assess, and answer test questions.

A large part of the guide is devoted to showing you what content to expect on the exam and to helping you better understand that content. In this guide are practice test questions so that you can see how well you have grasped the content. Then, answer explanations are provided so that you can understand why you missed certain questions.

Don't try to cram the night before you take your exam. This is not a wise strategy for a few reasons. First, your retention of the information will be low. Your time would be better used by reviewing information you already know rather than trying to learn a lot of new information. Second, you will likely become stressed as you try to gain a large amount of knowledge in a short amount of time. Third, you will be depriving yourself of sleep. So be sure to go to bed at a reasonable time the night before. Being well-rested helps you focus and remain calm.

Be sure to eat a substantial breakfast the morning of the exam. If you are taking the exam in the afternoon, be sure to have a good lunch as well. Being hungry is distracting and can make it difficult to focus. You have hopefully spent lots of time preparing for the exam. Don't let an empty stomach get in the way of success!

When travelling to the testing center, leave earlier than needed. That way, you have a buffer in case you experience any delays. This will help you remain calm and will keep you from missing your appointment time at the testing center.

Be sure to pace yourself during the exam. Don't try to rush through the exam. There is no need to risk performing poorly on the exam just so you can leave the testing center early. Allow yourself to use all of the allotted time if needed.

Remain positive while taking the exam even if you feel like you are performing poorly. Thinking about the content you should have mastered will not help you perform better on the exam.

Once the exam is complete, take some time to relax. Even if you feel that you need to take the exam again, you will be well served by some down time before you begin studying again. It's often easier to convince yourself to study if you know that it will come with a reward!

Test-Taking Strategies

1. Predicting the Answer

When you feel confident in your preparation for a multiple-choice test, try predicting the answer before reading the answer choices. This is especially useful on questions that test objective factual knowledge. By predicting the answer before reading the available choices, you eliminate the possibility that you will be distracted or led astray by an incorrect answer choice. You will feel more confident in your selection if you read the question, predict the answer, and then find your prediction among the answer choices. After using this strategy, be sure to still read all of the answer choices carefully and completely. If you feel unprepared, you should not attempt to predict the answers. This would be a waste of time and an opportunity for your mind to wander in the wrong direction.

2. Reading the Whole Question

Too often, test takers scan a multiple-choice question, recognize a few familiar words, and immediately jump to the answer choices. Test authors are aware of this common impatience, and they will sometimes prey upon it. For instance, a test author might subtly turn the question into a negative, or he or she might redirect the focus of the question right at the end. The only way to avoid falling into these traps is to read the entirety of the question carefully before reading the answer choices.

3. Looking for Wrong Answers

Long and complicated multiple-choice questions can be intimidating. One way to simplify a difficult multiple-choice question is to eliminate all of the answer choices that are clearly wrong. In most sets of answers, there will be at least one selection that can be dismissed right away. If the test is administered on paper, the test taker could draw a line through it to indicate that it may be ignored; otherwise, the test taker will have to perform this operation mentally or on scratch paper. In either case, once the obviously incorrect answers have been eliminated, the remaining choices may be considered. Sometimes identifying the clearly wrong answers will give the test taker some information about the correct answer. For instance, if one of the remaining answer choices is a direct opposite of one of the eliminated answer choices, it may well be the correct answer. The opposite of obviously wrong is obviously right! Of course, this is not always the case. Some answers are obviously incorrect simply because they are irrelevant to the question being asked. Still, identifying and eliminating some incorrect answer choices is a good way to simplify a multiple-choice question.

4. Don't Overanalyze

Anxious test takers often overanalyze questions. When you are nervous, your brain will often run wild, causing you to make associations and discover clues that don't actually exist. If you feel that this may be a problem for you, do whatever you can to slow down during the test. Try taking a deep breath or counting to ten. As you read and consider the question, restrict yourself to the particular words used by the author. Avoid thought tangents about what the author *really* meant, or what he or she was *trying* to say. The only things that matter on a multiple-choice test are the words that are actually in the question. You must avoid reading too much into a multiple-choice question, or supposing that the writer meant something other than what he or she wrote.

5. No Need for Panic

It is wise to learn as many strategies as possible before taking a multiple-choice test, but it is likely that you will come across a few questions for which you simply don't know the answer. In this situation, avoid panicking. Because most multiple-choice tests include dozens of questions, the relative value of a single wrong answer is small. As much as possible, you should compartmentalize each question on a multiple-choice test. In other words, you should not allow your feelings about one question to affect your success on the others. When you find a question that you either don't understand or don't know how to answer, just take a deep breath and do your best. Read the entire question slowly and carefully. Try rephrasing the question a couple of different ways. Then, read all of the answer choices carefully. After eliminating obviously wrong answers, make a selection and move on to the next question.

6. Confusing Answer Choices

When working on a difficult multiple-choice question, there may be a tendency to focus on the answer choices that are the easiest to understand. Many people, whether consciously or not, gravitate to the answer choices that require the least concentration, knowledge, and memory. This is a mistake. When you come across an answer choice that is confusing, you should give it extra attention. A question might be confusing because you do not know the subject matter to which it refers. If this is the case, don't eliminate the answer before you have affirmatively settled on another. When you come across an answer choice of this type, set it aside as you look at the remaining choices. If you can confidently assert that one of the other choices is correct, you can leave the confusing answer aside. Otherwise, you will need to take a moment to try to better understand the confusing answer choice. Rephrasing is one way to tease out the sense of a confusing answer choice.

7. Your First Instinct

Many people struggle with multiple-choice tests because they overthink the questions. If you have studied sufficiently for the test, you should be prepared to trust your first instinct once you have carefully and completely read the question and all of the answer choices. There is a great deal of research suggesting that the mind can come to the correct conclusion very quickly once it has obtained all of the relevant information. At times, it may seem to you as if your intuition is working faster even than your reasoning mind. This may in fact be true. The knowledge you obtain while studying may be retrieved from your subconscious before you have a chance to work out the associations that support it. Verify your instinct by working out the reasons that it should be trusted.

8. Key Words

Many test takers struggle with multiple-choice questions because they have poor reading comprehension skills. Quickly reading and understanding a multiple-choice question requires a mixture of skill and experience. To help with this, try jotting down a few key words and phrases on a piece of scrap paper. Doing this concentrates the process of reading and forces the mind to weigh the relative importance of the question's parts. In selecting words and phrases to write down, the test taker thinks about the question more deeply and carefully. This is especially true for multiple-choice questions that are preceded by a long prompt.

9. Subtle Negatives

One of the oldest tricks in the multiple-choice test writer's book is to subtly reverse the meaning of a question with a word like *not* or *except*. If you are not paying attention to each word in the question, you can easily be led astray by this trick. For instance, a common question format is, "Which of the following is...?" Obviously, if the question instead is, "Which of the following is not...?," then the answer will be quite different. Even worse, the test makers are aware of the potential for this mistake and will include one answer choice that would be correct if the question were not negated or reversed. A test taker who misses the reversal will find what he or she believes to be a correct answer and will be so confident that he or she will fail to reread the question and discover the original error. The only way to avoid this is to practice a wide variety of multiple-choice questions and to pay close attention to each and every word.

10. Reading Every Answer Choice

It may seem obvious, but you should always read every one of the answer choices! Too many test takers fall into the habit of scanning the question and assuming that they understand the question because they recognize a few key words. From there, they pick the first answer choice that answers the question they believe they have read. Test takers who read all of the answer choices might discover that one of the latter answer choices is actually *more* correct. Moreover, reading all of the answer choices can remind you of facts related to the question that can help you arrive at the correct answer. Sometimes, a misstatement or incorrect detail in one of the latter answer choices will trigger your memory of the subject and will enable you to find the right answer. Failing to read all of the answer choices is like not reading all of the items on a restaurant menu: you might miss out on the perfect choice.

11. Spot the Hedges

One of the keys to success on multiple-choice tests is paying close attention to every word. This is never truer than with words like almost, most, some, and sometimes. These words are called "hedges" because they indicate that a statement is not totally true or not true in every place and time. An absolute statement will contain no hedges, but in many subjects, the answers are not always straightforward or absolute. There are always exceptions to the rules in these subjects. For this reason, you should favor those multiple-choice questions that contain hedging language. The presence of qualifying words indicates that the author is taking special care with his or her words, which is certainly important when composing the right answer. After all, there are many ways to be wrong, but there is only one way to be right! For this reason, it is wise to avoid answers that are absolute when taking a multiple-choice test. An absolute answer is one that says things are either all one way or all another. They often include words like *every, always, best*, and *never*. If you are taking a multiple-choice test in a subject that doesn't lend itself to absolute answers, be on your guard if you see any of these words.

12. Long Answers

In many subject areas, the answers are not simple. As already mentioned, the right answer often requires hedges. Another common feature of the answers to a complex or subjective question are qualifying clauses, which are groups of words that subtly modify the meaning of the sentence. If the question or answer choice describes a rule to which there are exceptions or the subject matter is complicated, ambiguous, or confusing, the correct answer will require many words in order to be expressed clearly and accurately. In essence, you should not be deterred by answer choices that seem excessively long. Oftentimes, the author of the text will not be able to write the correct answer without offering some qualifications and modifications. Your job is to read the answer choices thoroughly and completely and to select the one that most accurately and precisely answers the question.

13. Restating to Understand

Sometimes, a question on a multiple-choice test is difficult not because of what it asks but because of how it is written. If this is the case, restate the question or answer choice in different words. This process serves a couple of important purposes. First, it forces you to concentrate on the core of the question. In order to rephrase the question accurately, you have to understand it well. Rephrasing the question will concentrate your mind on the key words and ideas. Second, it will present the information to your mind in a fresh way. This process may trigger your memory and render some useful scrap of information picked up while studying.

14. True Statements

Sometimes an answer choice will be true in itself, but it does not answer the question. This is one of the main reasons why it is essential to read the question carefully and completely before proceeding to the answer choices. Too often, test takers skip ahead to the answer choices and look for true statements. Having found one of these, they are content to select it without reference to the question above. Obviously, this provides an easy way for test makers to play tricks. The savvy test taker will always read the entire question before turning to the answer choices. Then, having settled on a correct answer choice, he or she will refer to the original question and ensure that the selected answer is relevant. The mistake of choosing a correct-but-irrelevant answer choice is especially common on questions related to specific pieces of objective knowledge. A prepared test taker will have a wealth of factual knowledge at his or her disposal, and should not be careless in its application.

15. No Patterns

One of the more dangerous ideas that circulates about multiple-choice tests is that the correct answers tend to fall into patterns. These erroneous ideas range from a belief that B and C are the most common right answers, to the idea that an unprepared test-taker should answer "A-B-A-C-A-D-A-B-A." It cannot be emphasized enough that pattern-seeking of this type is exactly the WRONG way to approach a multiple-choice test. To begin with, it is highly unlikely that the test maker will plot the correct answers according to some predetermined pattern. The questions are scrambled and delivered in a random order. Furthermore, even if the test maker was following a pattern in the assignation of correct answers, there is no reason why the test taker would know which pattern he or she was using. Any attempt to discern a pattern in the answer choices is a waste of time and a distraction from the real work of taking the test. A test taker would be much better served by extra preparation before the test than by reliance on a pattern in the answers.

FREE DVD OFFER

Don't forget that doing well on your exam includes both understanding the test content and understanding how to use what you know to do well on the test. We offer a completely FREE Test Taking Tips DVD that covers world class test taking tips that you can use to be even more successful when you are taking your test.

All that we ask is that you email us your feedback about your study guide. To get your **FREE Test Taking Tips DVD**, email freedvd@studyguideteam.com with "FREE DVD" in the subject line and the following information in the body of the email:

- The title of your study guide.
- Your product rating on a scale of 1-5, with 5 being the highest rating.
- Your feedback about the study guide. What did you think of it?
- Your full name and shipping address to send your free DVD.

Introduction to the Upper Level SSAT

Function of the Test

The Secondary School Admission Test (SSAT) is a standardized test used for students that are applying to an independent or private school. The Upper Level SSAT is administered to students currently in grades 8 through 11, applying to grades 9 through 12 and beyond to post-graduate opportunities, and it evaluates math, reading, and verbal skills. The test is used in the United States and is also available in several other countries throughout the world to determine if students have the necessary skills for success in a college preparatory program.

The SSAT is administered in a standard testing format and a Flex testing format. The standard SSAT is administered on 8 specific dates throughout the year, while the Flex test is administered to a student on any other date than the standard test. The Flex test is administered by an educational consultant or school at the student's request.

Test Administration

The standard SSAT is offered on eight Saturdays throughout the year. The standard test is available at hundreds of testing centers in the US and locations throughout the world. Sunday testing is available for religious reasons, but must be approved before registration. Students may repeat the standard test without penalty. Students must create an account on the SSAT website in order to register. This account also allows students to print their admission tickets and receive their test scores. Registration opens about 10 weeks prior to a testing date. Late registration begins 3 weeks before the test date and rush registration starts 10 days before the test date. Late and rush registrations incur additional fees.

Students who cannot attend the standard test dates can opt to take the Flex test. The Flex test may be provided in an open or closed format. A school may administer an open Flex test on a date other than the standard test for all registrants. The closed format is administered in a small group or individually at as school with an educational consultant. Students may only take the Flex SSAT once in an academic year.

Testing accommodations are available for students with disabilities. Students requiring accommodations must apply and be approved before registering for the test. Approval is only required once in an academic year.

Test Format

The Upper Level SSAT consists of multiple-choice questions in Quantitative (Math), Verbal, Reading Comprehension, and Experimental sections, and a writing sample. The writing sample is not scored, but is sent to prospective schools to demonstrate a student's writing ability.

The writing sample asks the student to choose between two writing prompts—a traditional essay or a creative writing stimulus—and to write an original essay or story based on the prompt. The Quantitative section is broken down into two parts, each with 25 questions in the areas of number concepts, geometry, algebra, and probability. A calculator is not permitted. The Verbal section includes 60 questions: 30 synonyms and 30 analogies. The Reading Comprehension section includes a total of 40 questions relating to several short passages that range from approximately 250-350 words. The passages may be pulled from literary fiction, social sciences, humanities, and sciences and the questions focus mainly on reading comprehension skills, including inferences, identifying the main idea, and determining the author's purpose. The Experimental section includes 16 questions (6 Verbal, 5 Reading, and 5 Quantitative) in order to test their reliability for SSAT forms to be used in the future. The Experimental section is not scored. The testing period and sections are broken down as follows:

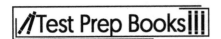

Section	Number of Questions	Question Type	Time Allotted
Writing Sample	1	Written Response	25 minutes
Break			5 minutes
Quantitative (Part 1)	25	Multiple Choice	30 minutes
Reading	40	Multiple Choice	40 minutes
Break			10 minutes
Verbal	60	Multiple Choice	30 minutes
Quantitative (Part 2)	25	Multiple Choice	30 minutes
Experimental	16	Multiple Choice	15 minutes

Scoring

In the Upper Level SSAT, students are given one point for each correct answer, and they lose a quarter of a point for each incorrect answer. Unanswered questions do not affect the score either way.

A free scoring report is available online through a student's SSAT account roughly 2 weeks after the test date. The report includes a narrative explanation of the scores, along with a raw score, a scaled score, percentile rank, and total scaled score. The possible scaled score ranges from 440 – 710 per section, and the percentile rank is from 1 – 99. The total scaled score range is 1320 – 2130, with a mid-point of 1725.

The SSAT is a norm-referenced test, meaning that the score is compared to a norm group of test takers' scores from the last 3 years. The score report includes the student's scores, as well as the average norm scores in each section for comparison. The percentile rank shows how a student performed relative to the norm group. For example, if a student's percentile is 80, it means he or she scored the same or better than 80% of those in the norm group. Schools use many factors to select students for admission, so a "good score" or passing score is difficult to determine.

Recent/Future Developments

The SSAT testing accommodations were updated in the 2016-2017 academic year. The official guidelines for accommodations can be accessed on the SSAT.org website.

Quantitative Reasoning

Number Concepts and Operations

The Position of Numbers Relative to Each Other

Place Value of a Digit

Numbers count in groups of 10. This means that the number in the 10's place will remain the same throughout the set of natural and whole numbers. This is referred to as working within a base 10 numeration system. Only the numbers from zero to nine are used to represent any number. The foundation for doing this involves **place value**. Numbers are written side by side. This is to show the amount in each place value.

For place value, let's look at how the number 10 is different from the numbers zero to 9. Ten has two digits instead of just one. The one is in the tens' place, and the zero is in the ones' place. Therefore, there is one group of tens and zero ones. 11 has one 10 and one 1. Considering numbers from 11 to 19 should be the next step in understanding place value. Each value within this range of numbers consists of one group of 10 and a specific number of leftover ones. Counting by tens can be practiced once the tens column is understood. This process consists of increasing the number in the tens' place by one. For example, counting by 10 starting at 17 would result in the next four values being 27, 37, 47, and 57.

A place value chart can be used for understanding and learning about numbers that have more digits. Here is an example of a place value chart:

	MILLIONS			THOUSANDS			ONES			.	DECIMALS		
billions	hundred millions	ten millions	millions	hundred thousands	ten thousands	thousands	hundreds	tens	ones		tenths	hundredths	thousandths

In the number 1,234, there are 4 ones and 3 tens. The 2 is in the hundreds' place, and the one is in the thousands' place. Note that each group of three digits is separated by a comma. The 2 has a value that is 10 times greater than the 3. Every place to the left has a value 10 times greater than the place to its right. Also, each group of three digits is also known as a **period**. 234 is in the ones' period.

The number 1,234 can be written out as *one-thousand, two hundred thirty-four*. The process of writing out numbers is known as the **decimal system**. It is also based on groups of 10. The place value chart is a helpful tool in using this system. In order to write out a number, it always starts with the digit(s) in the highest period. For example, in the number 23,815,467, the 23 is in highest place and is in the millions' period. The number is read *twenty-three million, eight hundred fifteen thousand, four hundred sixty-seven*. Each period is written separately through the use of commas. Also, no "ands" are used within the number. Another way to think about the number 23,815,467 is through the use of an addition problem. For example,

$$23{,}815{,}467 = 20{,}000{,}000 + 3{,}000{,}000 + 800{,}000 + 10{,}000 + 5{,}000 + 400 + 60 + 7$$

This expression is known as **expanded form**. The actual number 23,815,467 is considered to be in **standard form**.

In order to compare whole numbers with many digits, place value can be used. In each number to be compared, it is necessary to find the highest place value in which the numbers differ and to compare the value within that place value. For example,

$$4{,}523{,}345 < 4{,}532{,}456$$

because of the values in the ten thousands' place. A similar process can be used for decimals. However, number lines can also be used. Tick marks can be placed between two whole numbers on the number line that represent tenths, hundredths, etc. Each number being compared can then be plotted. The value farthest to the right on the number line is the largest.

Classifying Real Numbers

The mathematical number system is made up of two general types of numbers: real and complex. **Real numbers** are those that are used in normal settings, while **complex numbers** are those composed of both a real number and an imaginary one. Imaginary numbers are the result of taking the square root of -1, and $\sqrt{-1} = i$.

The real number system is often explained using a Venn diagram similar to the one below. After a number has been labeled as a real number, further classification occurs when considering the other groups in this diagram. If a number is a never-ending, non-repeating decimal, it falls in the **irrational** category. Otherwise, it is **rational**. Furthermore, if a number does not have a fractional part, it is classified as an **integer**, such as -2, 75, or 0. **Whole numbers** are an even smaller group that only includes positive integers and 0. The last group, **natural numbers**, is made up of only positive integers, such as 2, 56, or 12.

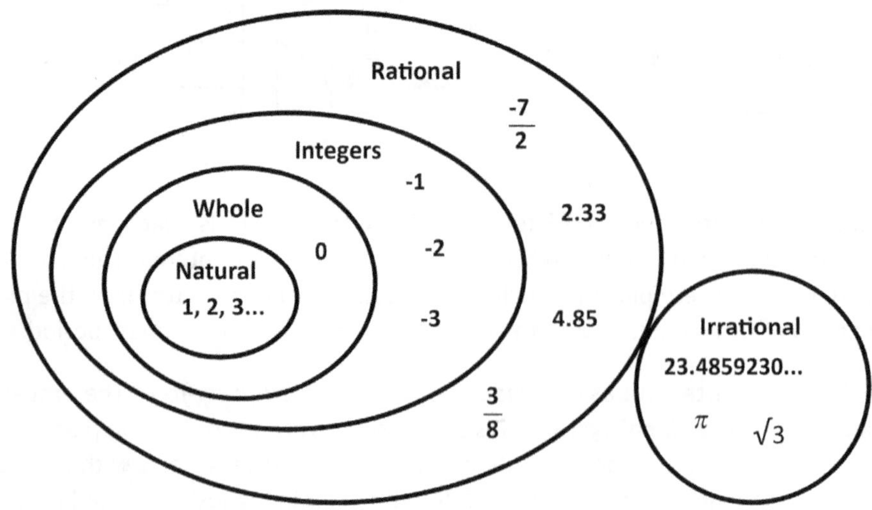

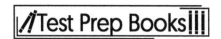

Integers are the whole numbers together with their negatives. They include numbers like 5, -24, 0, -6, and 15. They do not include fractions or numbers that have digits after the decimal point.

Rational numbers are all numbers that can be written as a fraction using integers. They are inclusive of integers, fractions, and decimals that terminate, or end (1.7, 0.04213) or repeat ($0.136\overline{5}$). A **fraction** is written as $\frac{x}{y}$ and represents the quotient of *x* being divided by *y*. More practically, it means dividing the whole into *y* equal parts, then taking *x* of those parts.

Examples of rational numbers include $\frac{1}{2}$ and $\frac{5}{4}$. The number on the top is called the **numerator,** and the number on the bottom is called the **denominator**. Because every integer can be written as a fraction with a denominator of 1, (e.g. $\frac{3}{1} = 3$), every integer is also a rational number.

A **number line** typically consists of integers (...3, 2, 1, 0, -1, -2, -3...), and is used to visually represent the value of a rational number. Each rational number has a distinct position on the line determined by comparing its value with the displayed values on the line. For example, if plotting -1.5 on the number line below, it is necessary to recognize that the value of -1.5 is .5 less than -1 and .5 greater than -2. Therefore, -1.5 is plotted halfway between -1 and -2.

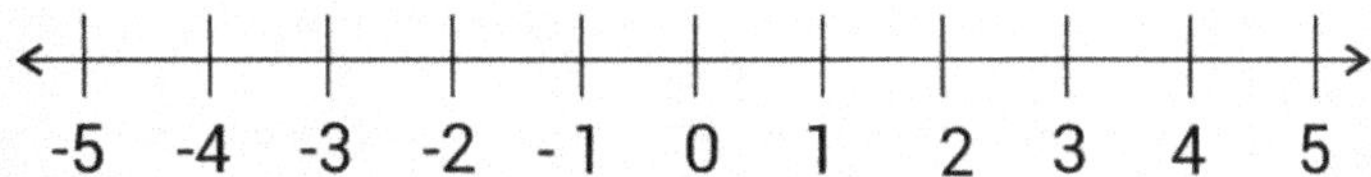

The number system that is used consists of only ten different digits or characters. However, this system is used to represent an infinite number of values. As mentioned, the **place value system** makes this infinite number of values possible. The position in which a digit is written corresponds to a given value. Starting from the decimal point (which is implied, if not physically present), each subsequent place value to the left represents a value greater than the one before it. Conversely, starting from the decimal point, each subsequent place value to the right represents a value less than the one before it.

In accordance with the **base-10 system**, the value of a digit increases by a factor of ten each place it moves to the left. For example, consider the number 7. Moving the digit one place to the left (70), increases its value by a factor of 10 ($7 \times 10 = 70$). Moving the digit two places to the left (700) increases its value by a factor of 10 twice ($7 \times 10 \times 10 = 700$). Moving the digit three places to the left (7,000) increases its value by a factor of 10 three times ($7 \times 10 \times 10 \times 10 = 7{,}000$), and so on.

Conversely, the value of a digit decreases by a factor of ten each place it moves to the right. (Note that multiplying by $\frac{1}{10}$ is equivalent to dividing by 10). For example, consider the number 40. Moving the digit one place to the right (4) decreases its value by a factor of 10 ($40 \div 10 = 4$). Moving the digit two places to the right (0.4), decreases its value by a factor of 10 twice ($40 \div 10 \div 10 = 0.4$) or ($40 \times \frac{1}{10} \times \frac{1}{10} = 0.4$). Moving the digit three places to the right (0.04) decreases its value by a factor of 10 three times:

$$40 \div 10 \div 10 \div 10 = 0.04 \text{ or}$$

$$(40 \times \frac{1}{10} \times \frac{1}{10} \times \frac{1}{10} = 0.04)$$

and so on.

Prime and Composite Numbers

Whole numbers are classified as either prime or composite. A **prime number** can only be divided evenly by itself and one. For example, the number 11 can only be divided evenly by 11 and one; therefore, 11 is a prime number. A helpful way to visualize a prime number is to use concrete objects and try to divide them into equal piles. If dividing 11 coins, the only way to divide them into equal piles is to create 1 pile of 11 coins or to create 11 piles of 1 coin each. Other examples of prime numbers include 2, 3, 5, 7, 13, 17, and 19.

A **composite number** is any whole number that is not a prime number. A composite number is a number that can be divided evenly by one or more numbers other than itself and one. For example, the number 6 can be divided evenly by 2 and 3. Therefore, 6 is a composite number. If dividing 6 coins into equal piles, the possibilities are 1 pile of 6 coins, 2 piles of 3 coins, 3 piles of 2 coins, or 6 piles of 1 coin. Other examples of composite numbers include 4, 8, 9, 10, 12, 14, 15, 16, 18, and 20.

To determine if a number is a prime or composite number, the number is divided by every whole number greater than one and less than its own value. If it divides evenly by any of these numbers, then the number is composite. If it does not divide evenly by any of these numbers, then the number is prime. For example, 5 cannot be divided evenly by 2, 3, or 4. Therefore, 5 must be a prime number.

Ordering Numbers

A common question type asks to order rational numbers from least to greatest or greatest to least. The numbers will come in a variety of formats, including decimals, percentages, roots, fractions, and whole numbers. These questions test for knowledge of different types of numbers and the ability to determine their respective values.

Before discussing ordering all numbers, let's start with decimals.

To compare decimals and order them by their value, utilize a method similar to that of ordering large numbers.

The main difference is where the comparison will start. Assuming that any numbers to left of the decimal point are equal, the next numbers to be compared are those immediately to the right of the decimal point. If those are equal, then move on to compare the values in the next decimal place to the right.

For example:

Which number is greater, 12.35 or 12.38?

Check that the values to the left of the decimal point are equal:

$$12 = 12$$

Next, compare the values of the decimal place to the right of the decimal:

$$12.3 = 12.3$$

Those are also equal in value.

Finally, compare the value of the numbers in the next decimal place to the right on both numbers:

12.3**5** and 12.3**8**

Here the 5 is less than the 8, so the final way to express this inequality is:

$$12.35 < 12.38$$

Comparing decimals is regularly exemplified with money because the "cents" portion of money ends in the hundredths' place. When paying for gasoline or meals in restaurants, and even in bank accounts, if enough errors are made when calculating numbers to the hundredths place, they can add up to dollars and larger amounts of money over time.

Now that decimal ordering has been explained, let's expand and consider all real numbers. Whether the question asks to order the numbers from greatest to least or least to greatest, the crux of the question is the same—convert the numbers into a common format. Generally, it's easiest to write the numbers as whole numbers and decimals so they can be placed on a number line. Follow these examples to understand this strategy.

1) Order the following rational numbers from greatest to least:

$$\sqrt{36}, 0.65, 78\%, \frac{3}{4}, 7, 90\%, \frac{5}{2}$$

Of the seven numbers, the whole number (7) and decimal (0.65) are already in an accessible form, so concentrate on the other five.

First, the square root of 36 equals 6. (If the test asks for the root of a non-perfect root, determine which two whole numbers the root lies between.) Next, convert the percentages to decimals. A percentage means "per hundred," so this conversion requires moving the decimal point two places to the left, leaving 0.78 and 0.9.

Lastly, evaluate the fractions:

$$\frac{3}{4} = \frac{75}{100} = 0.75\,; \frac{5}{2} = 2\frac{1}{2} = 2.5$$

Now, the only step left is to list the numbers in the requested order:

$$7, \sqrt{36}, \frac{5}{2}, 90\%, 78\%, \frac{3}{4}, 0.65$$

2) Order the following rational numbers from least to greatest:

$$2.5, \sqrt{9}, -10.5, 0.853, 175\%, \sqrt{4}, \frac{4}{5}$$

$$\sqrt{9} = 3$$

$$175\% = 1.75$$

$$\sqrt{4} = 2$$

$$\frac{4}{5} = 0.8$$

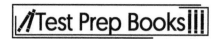

From least to greatest, the answer is:

$$-10.5, \frac{4}{5}, 0.853, 175\%, \sqrt{4}, 2.5, \sqrt{9}$$

When comparing or ordering numbers, the numbers should be written in the same format (decimal or fraction), if possible. For example, $\sqrt{49}$, 7.3, and $\frac{15}{2}$ are easier to order if each one is converted to a decimal, such as 7, 7.3, and 7.5 (converting fractions and decimals is covered in a subsequent section). A number line is used to order and compare the numbers. Any number that is to the right of another number is greater than that number. Conversely, a number positioned to the left of a given number is less than that number.

Basic Addition, Subtraction, Multiplication, and Division

Gaining more of something is related to addition, while taking something away relates to subtraction. Vocabulary words such as *total, more, less, left*, and *remain* are common when working with these problems. The $+$ sign means *plus*. This shows that addition is happening. The $-$ sign means *minus*. This shows that subtraction is happening. The symbols will be important when you write out equations.

Addition

Addition can also be defined in equation form. For example, $4 + 5 = 9$ shows that $4 + 5$ is the same as 9. Therefore, $9 = 9$, and "four plus five equals nine." When two quantities are being added together, the result is called the **sum**. Therefore, the sum of 4 and 5 is 9. The numbers being added, such as 4 and 5, are known as the **addends.**

Subtraction

Subtraction can also be in equation form. For example, $9 - 5 = 4$ shows that $9 - 5$ is the same as 4 and that "9 minus 5 is 4." The result of subtraction is known as a **difference.** The difference of $9 - 5$ is 4. 4 represents the amount that is left once the subtraction is done. The order in which subtraction is completed does matter. For example, $9 - 5$ and $5 - 9$ do not result in the same answer. $5 - 9$ results in a negative number. So, subtraction does not adhere to the commutative or associative property. The order in which subtraction is completed is important.

Multiplication

Multiplication is when we add equal amounts. The answer to a multiplication problem is called a **product**. Products stand for the total number of items within different groups. The symbol for multiplication is $\times$ or $\cdot$. We say 2×3 or $2 \cdot 3$ means "2 times 3."

As an example, there are three sets of four apples. The goal is to know how many apples there are in total. Three sets of four apples gives:

$$4 + 4 + 4 = 12$$

Also, three times four apples gives $3 \times 4 = 12$. Therefore, for any whole numbers *a* and *b,* where *a* is not equal to zero, $a \times b = b + b + \cdots b$, where *b* is added *a* times. Also, $a \times b$ can be thought of as the number of units in a rectangular block consisting of *a* rows and *b* columns. For example, 3×7 is equal to the number of squares in the following rectangle:

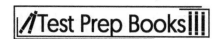

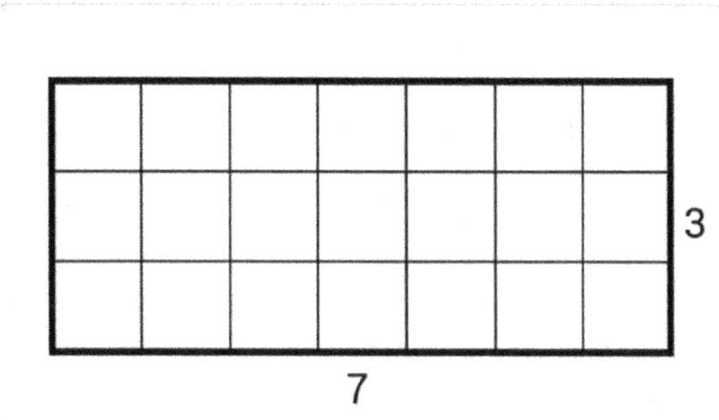

The answer is 21, and there are 21 squares in the rectangle.

When any number is multiplied by one (for example, $8 \times 1 = 8$), the value of original number does not change. Therefore, 1 is the **multiplicative identity**. For any whole number *a*, $1 \times a = a$. Also, any number multiplied by zero results in zero. Therefore, for any whole number *a*, $0 \times a = 0$.

Another method of multiplication can be done with the use of an **area model**. An area model is a rectangle that is divided into rows and columns that match up to the number of place values within each number. For example, $29 \times 65 = 25 + 4$ and $66 = 60 + 5$. The products of those 4 numbers are found within the rectangle and then summed up to get the answer. The entire process is:

$$(60 \times 25) + (5 \times 25) + (60 \times 4) + (5 \times 4)$$

$$1{,}500 + 240 + 125 + 20$$

$$1{,}885$$

Here is the actual area model:

	25	4
60	60x25 1,500	60x4 240
5	5x25 125	5x4 20

$$\begin{array}{r} 1{,}500 \\ 240 \\ 125 \\ +\quad 20 \\ \hline 1{,}885 \end{array}$$

Division

Division is based on dividing a given number into parts. The simplest problem involves dividing a number into equal parts. For example, if a pack of 20 pencils is to be divided among 10 children, you would have to divide 20 by 10. In this example, each child would receive 2 pencils.

The symbol for division is $\div$ or /. The equation above is written as

$$20 \div 10 = 2, \text{ or } 20 / 10 = 2$$

This means "20 divided by 10 is equal to 2." Division can be explained as the following: for any whole numbers *a* and *b*, where *b* is not equal to zero, $a \div b = c$ if—and only if—$a = b \times c$. This means, division can be thought of as a multiplication problem with a missing part. For instance, calculating $20 \div 10$ is the same as asking the following: "If there are 20 items in total with 10 in each group, how many are in each group?" Therefore, 20 is equal to ten times what value? This question is the same as asking, "If there are 20 items in total with 2 in each group, how many groups are there?" The answer to each question is 2.

In a division problem, *a* is known as the **dividend,** *b* is the **divisor**, and *c* is the **quotient**. Zero cannot be divided into parts. Therefore, for any nonzero whole number *a,* $0 \div a = 0$. Also, division by zero is undefined. Dividing an amount into zero parts is not possible.

More difficult division problems involve dividing a number into equal parts, but having some left over. An example is dividing a pack of 20 pencils among 8 friends so that each friend receives the same number of pencils. In this setting, each friend receives 2 pencils, but there are 4 pencils leftover. 20 is the dividend, 8 is the divisor, 2 is the quotient, and 4 is known as the **remainder**. Within this type of division problem, for whole numbers *a, b, c*, and *d*, $a \div b = c$ with a remainder of *d.* This is true if and only if $a = (b \times c) + d$. When calculating $a \div b$, if there is no remainder, *a* is said to be *divisible* by *b.* **Even numbers** are all divisible by the number 2. **Odd numbers** are not divisible by 2. An odd number of items cannot be paired up into groups of 2 without having one item leftover.

Dividing a number by a single digit or two digits can be turned into repeated subtraction problems. An area model can be used throughout the problem that represents multiples of the divisor. For example, the answer to $8580 \div 55$ can be found by subtracting 55 from 8580 one at a time and counting the total number of subtractions necessary.

However, a simpler process involves using larger multiples of 55. First,

$$100 \times 55 = 5{,}500$$

is subtracted from 8,580, and 3,080 is leftover. Next,

$$50 \times 55 = 2{,}750$$

is subtracted from 3,080 to obtain 380.

$$5 \times 55 = 275$$

is subtracted from 330 to obtain 55, and finally:

$$1 \times 55 = 55$$

is subtracted from 55 to obtain zero. Therefore, there is no remainder, and the answer is:

$$100 + 50 + 5 + 1 = 156.$$

Here is a picture of the area model and the repeated subtraction process:

8580÷55

	55
100	5500
50	2750
5	275
1	55

```
55 | 8580
    -5500  (100 x 55)
     3080
    -2750  (50 x 55)
      330
     -275  (5 x 55)
       55
      -55  (1 x 55)
        0
```

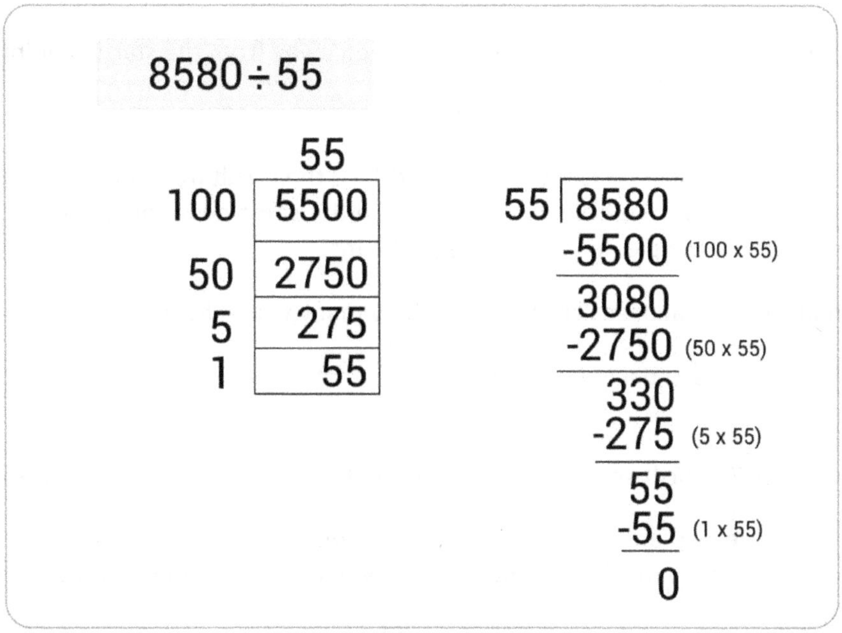

If you want to check the answer of a division problem, multiply the answer by the divisor. This will help you check to see if the dividend is obtained. If there is a remainder, the same process is done, but the remainder is added on at the end to try to match the dividend. In the previous example, $156 \times 64 = 9984$ would be the checking procedure. Dividing decimals involves the same repeated subtraction process. The only difference would be that the subtractions would involve numbers that include values in the decimal places. Lining up decimal places is crucial in this type of problem.

Using Operations in Math and Real-World Problems

Addition and subtraction are "inverse operations." Adding a number and then subtracting the same number will cancel each other out. This results in the original number, and vice versa. For example:

$$8 + 7 - 7 = 8 \text{ and}$$

$$137 - 100 + 100 = 137$$

Multiplication and division are also **inverse operations**. So, multiplying by a number and then dividing by the same number results in the original number. For example:

$$8 \times 2 \div 2 = 8 \text{ and } 12 \div 4 \times 4 = 12$$

Inverse operations are used to work backwards to solve problems. In the case that 7 and a number add to 18, the inverse operation of subtraction is used to find the unknown value ($18 - 7 = 11$). If a school's entire 4th grade was divided evenly into 3 classes each with 22 students, the inverse operation of multiplication is used to determine the total students in the grade ($22 \times 3 = 66$). More scenarios involving inverse operations are listed in the tables below.

Word problems take concepts you learned in the classroom and turn them into real-life situations. Some parts of the problem are known and at least one part is unknown. There are three types of instances in which something can be unknown: the starting point, the change, or the final result. Any of these can be missing from the information they give you.

For an addition problem, the change is the quantity of a new amount added to the starting point.

For a subtraction problem, the change is the quantity taken away from the starting point.

Regarding addition, the given equation is $3 + 7 = 10$.

The number 3 is the starting point. 7 is the change, and 10 is the result from adding a new amount to the starting point. Different word problems can arise from this same equation, depending on which value is the unknown. For example, here are three problems:

- If a boy had 3 pencils and was given 7 more, how many would he have in total?
- If a boy had 3 pencils and a girl gave him more so that he had 10 in total, how did she give to him?
- A boy was given 7 pencils so that he had 10 in total. How many did he start with?

All three problems involve the same equation. Finding out which part of the equation is missing is the key to solving each word problem. The missing answers would be 10, 7, and 3, respectively.

In terms of subtraction, the same three scenarios can occur. Imagine the given equation is $6 - 4 = 2$.

The number 6 is the starting point, 4 is the change, and 2 is the new amount that is the result from taking away an amount from the starting point. Again, different types of word problems can arise from this equation. For example, here are three possible problems:

- If a girl had 6 quarters and 2 were taken away, how many would be left over?
- If a girl had 6 quarters, purchased a pencil, and had 2 quarters left over, how many did she pay with?
- If a girl paid for a pencil with 4 quarters and had 2 quarters left over, how many did she start with?

The three question types follow the structure of the addition word problems. Finding out whether the starting point, the change, or the final result is missing is the goal in solving the problem. The missing answers would be 2, 4, and 6, respectively.

The three addition problems and the three subtraction word problems can be solved by using a picture, a number line, or an algebraic equation. If an equation is used, a question mark can be used to show the number we don't know. For example, $6 - 4 = ?$ can be written to show that the missing value is the result. Using equation form shows us what part of the addition or subtraction problem is missing.

Key words within a multiplication problem involve *times, product, doubled,* and *tripled.* Key words within a division problem involve *split, quotient, divided, shared, groups*, and *half*. Like addition and subtraction, multiplication and division problems also have three different types of missing values.

Multiplication consists of a certain number of groups, with the same number of items within each group, and the total amount within all groups. Therefore, each one of these amounts can be the missing value.

For example, the given equation is $5 \times 3 = 15$.

5 and 3 are interchangeable, so either amount can be the number of groups or the number of items within each group. 15 is the total number of items. Again, different types of word problems can arise from this equation. For example, here are three problems:

- If a classroom is serving 5 different types of apples for lunch and has 3 apples of each type, how many total apples are there to give to the students?
- If a classroom has 15 apples with 5 different types, how many of each type are there?
- If a classroom has 15 apples total with 3 of each type, how many types are there to choose from?

Each question involves using the same equation to solve. It is important to decide which part of the equation is the missing value. The answers to the problems are 15, 3, and 5, respectively.

Similar to multiplication, division problems involve a total amount, a number of groups having the same amount, and a number of items within each group. The difference between multiplication and division is that the starting point in a division problem is the total amount. This then gets divided into equal amounts.

For example, the equation is $48 \div 8 = 6$.

48 is the total number of items, which is being divided into 8 different groups. In order to do so, 6 items go into each group. Also, 8 and 6 are interchangeable. So, the 48 items could be divided into 6 groups of 8 items each. Therefore, different types of word problems can arise from this equation. For example, here are three types of problems:

- A boy needs 48 pieces of chalk. If there are 8 pieces in each box, how many boxes should he buy?
- A boy has 48 pieces of chalk. If each box has 6 pieces in it, how many boxes did he buy?
- A boy has partitioned all of his chalk into 8 piles, with 6 pieces in each pile. How many pieces does he have in total?

Each one of these questions involves the same equation. The third question can easily utilize the multiplication equation $8 \times 6 = ?$ instead of division. The answers are 6, 8, and 48.

Order of Operations

When reviewing calculations consisting of more than one operation, the order in which the operations are performed affects the resulting answer. Consider $5 \times 2 + 7$. Performing multiplication then addition results in an answer of 17 because

$$(5 \times 2 = 10; 10 + 7 = 17)$$

However, if the problem is written $5 \times (2 + 7)$, the order of operations dictates that the operation inside the parenthesis must be performed first. The resulting answer is 45 because

$$(2 + 7 = 9, \text{then } 5 \times 9 = 45)$$

The order in which operations should be performed is remembered using the acronym PEMDAS. PEMDAS stands for parenthesis, exponents, multiplication/division, addition/subtraction. Multiplication and division are performed in the same step, working from left to right with whichever comes first. Addition and subtraction are performed in the same step, working from left to right with whichever comes first.

Consider the following example:

$$8 \div 4 + 8(7 - 7)$$

Performing the operation inside the parenthesis produces

$$8 \div 4 + 8(0) \text{ or } 8 \div 4 + 8 \times 0$$

There are no exponents, so multiplication and division are performed next from left to right resulting in:

$$2 + 8 \times 0$$

then $2 + 0$. Finally, addition and subtraction are performed to obtain an answer of 2. Now consider the following example:

$$6 \times 3 + 3^2 - 6$$

Parenthesis are not applicable. Exponents are evaluated first, which brings us to:

$$6 \times 3 + 9 - 6$$

Then multiplication/division forms:

$$18 + 9 - 6$$

At last, addition/subtraction leads to the final answer of 21.

Properties of Operations

Properties of operations exist that make calculations easier and solve problems for missing values. The following table summarizes commonly used properties of real numbers.

Property	**Addition**	**Multiplication**
Commutative	$a + b = b + a$	$a \times b = b \times a$
Associative	$(a + b) + c = a + (b + c)$	$(a \times b) \times c = a \times (bc)$
Identity	$a + 0 = a;\ 0 + a = a$	$a \times 1 = a;\ 1 \times a = a$
Inverse	$a + (-a) = 0$	$a \times \frac{1}{a} = 1;\ a \neq 0$
Distributive	$a(b + c) = ab + ac$	

The **cumulative property of addition** states that the order in which numbers are added does not change the sum. Similarly, the **commutative property of multiplication** states that the order in which numbers are multiplied does not change the product. The **associative property** of addition and multiplication state that the grouping of numbers being added or multiplied does not change the sum or product, respectively. The commutative and associative properties are useful for performing calculations. For example, $(47 + 25) + 3$ is equivalent to $(47 + 3) + 25$, which is easier to calculate.

The **identity property of addition** states that adding zero to any number does not change its value. The **identity property of multiplication** states that multiplying a number by 1 does not change its value. The **inverse property of addition** states that the sum of a number and its opposite equals zero. Opposites are numbers that are the same with different signs (ex. 5 and -5; $-\frac{1}{2}$ and $\frac{1}{2}$). The **inverse property of multiplication** states that the product of a number (other than 0) and its reciprocal equals 1. **Reciprocal numbers** have numerators and denominators that are inverted (ex. $\frac{2}{5}$ and $\frac{5}{2}$). Inverse properties are useful for canceling quantities to find missing values (see algebra content). For example, $a + 7 = 12$ is solved by adding the inverse of 7(-7) to both sides in order to isolate a.

The **distributive property** states that multiplying a sum (or difference) by a number produces the same result as multiplying each value in the sum (or difference) by the number and adding (or subtracting) the products. Consider the following scenario: You are buying three tickets for a baseball game. Each ticket costs \$18. You are also charged a fee of \$2 per ticket for purchasing the tickets online. The cost is calculated:

$$3 \times 18 + 3 \times 2$$

Using the distributive property, the cost can also be calculated $3(18 + 2)$.

Adding and Subtracting Positive and Negative Numbers

Some problems require adding positive and negative numbers or subtracting positive and negative numbers. Adding a negative number to a positive one can be thought of a reducing or subtracting from the positive number, and the result should be less than the original positive number. For example, adding 8 and -3 is the same is subtracting 3 from 8; the result is 5. This can be visualized by imagining that the positive number (8) represents 8 apples that a student has in her basket. The negative number (-3) indicates the number of apples she is in debt or owes to her friend. In order to pay off her debt and "settle the score," she essentially is in possession of three fewer apples than in her basket (8 – 3 = 5), so she actually has five apples that are hers to keep. Should the negative addend be of higher magnitude than the positive addend (for example -9 + 3), the result will be negative, but "less negative" or closer to zero than the large negative number. This is because adding a positive value, even if relatively smaller, to a negative value, reduces the magnitude of the negative in the total. Considering the apple example again, if the girl owed 9 apples to her friend (-9) but she picked 3 (+3) off a tree and gave them to her friend, she now would only owe him six apples (-6), which reduced her debt burden (her negative number of apples) by three.

Subtracting positive and negative numbers works the same way with one key distinction: subtracting a negative number from a negative number yields a "less negative" or more positive result because again, this can be considered as removing or alleviating some debt. For example, if the student with the apples owed 5 apples to her friend, she essentially has -5 applies. If her mom gives that friend 10 apples on behalf of the girl, she now has removed the need to pay back the 5 apples and surpassed neutral (no net apples owed) and now her friend owes *her* five apples (+5). Stated mathematically -5 – (-10) = +5.

When subtracting integers and negative rational numbers, one has to change the problem to adding the opposite and then apply the rules of addition.

- Subtracting two positive numbers is the same as adding one positive and one negative number.
  - For example, $4.9 - 7.1$ is the same as $4.9 + (-7.1)$. The solution is -2.2 since the absolute value of -7.1 is greater than 4.9. Another example is $8.5 - 6.4$ which is the same as $8.5 + (-6.4)$. The solution is 2.1 since the absolute value of 8.5 is greater than 6.4.

- Subtracting a positive number from a negative number results in negative value.
  - For example, $(-12) - 7$ is the same as $(-12) + (-7)$ with a solution of -19.

- Subtracting a negative number from a positive number results in a positive value.
  - For example, $12 - (-7)$ is the same as $12 + 7$ with a solution of 19.

- For multiplication and division of integers and rational numbers, if both numbers are positive or both numbers are negative, the result is a positive value.
  - For example, $(-1.7) \times (-4)$ has a solution of 6.8 since both numbers are negative values.

- If one number is positive and another number is negative, the result is a negative value.
  - For example, $(-15) \div 5$ has a solution of -3 since there is one negative number.

Adding one positive and one negative number requires taking the absolute values and finding the difference between them. Then, the sign of the number that has the higher absolute value for the final solution is used.

Operations with Fractions, Decimals, and Percentages

Fractions

A **fraction** is a part of something that is whole. Items such as apples can be cut into parts to help visualize fractions. If an apple is cut into 2 equal parts, each part represents ½ of the apple. If each half is then cut into two parts, the apple now is cut into quarters. Each piece now represents ¼ of the apple. In this example, each part is equal because they all have the same size. Geometric shapes, such as circles and squares, can also be utilized to help visualize the idea of fractions. For example, a circle can be drawn on the board and divided into 6 equal parts:

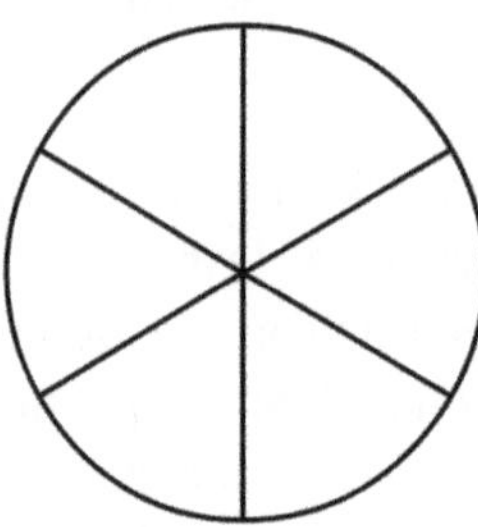

Shading can be used to represent parts of the circle that can be translated into fractions. The top of the fraction, the **numerator,** can represent how many segments are shaded. The bottom of the fraction, the **denominator,** can represent the number of segments that the circle is broken into. A pie is a good analogy to use in this example. If one piece of the circle is shaded, or one piece of pie is cut out, $^1/_6$ of the object is being referred to. An apple, a pie, or a circle can be utilized in order to compare simple fractions. For example, showing that ½ is larger than ¼ and that ¼ is smaller than $^1/_3$ can be accomplished through shading. A **unit fraction** is a fraction in which the numerator is 1, and the denominator is a positive whole number. It represents one part of a whole—one piece of pie.

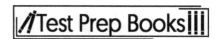

Imagine that an apple pie has been baked for a holiday party, and the full pie has eight slices. After the party, there are five slices left. How could the amount of the pie that remains be expressed as a fraction? The numerator is 5 since there are 5 pieces left, and the denominator is 8 since there were eight total slices in the whole pie. Thus, expressed as a fraction, the leftover pie totals $\frac{5}{8}$ of the original amount.

Fractions come in three different varieties: proper fractions, improper fractions, and mixed numbers. **Proper fractions** have a numerator less than the denominator, such as $\frac{3}{8}$, but **improper fractions** have a numerator greater than the denominator, such as $\frac{15}{8}$. **Mixed numbers** combine a whole number with a proper fraction, such as $3\frac{1}{2}$. Any mixed number can be written as an improper fraction by multiplying the integer by the denominator, adding the product to the value of the numerator, and dividing the sum by the original denominator. For example:

$$3\frac{1}{2} = \frac{3 \times 2 + 1}{2} = \frac{7}{2}$$

Whole numbers can also be converted into fractions by placing the whole number as the numerator and making the denominator 1. For example, $3 = \frac{3}{1}$.

The bar in a fraction represents division. Therefore $^6/_5$ is the same as $6 \div 5$. In order to rewrite it as a mixed number, division is performed to obtain $6 \div 5 = 1\ R1$. The remainder is then converted into fraction form. The actual remainder becomes the numerator of a fraction, and the divisor becomes the denominator. Therefore $1\ R1$ is written as $1\frac{1}{5}$, a mixed number. A mixed number can also decompose into the addition of a whole number and a fraction. For example:

$$1\frac{1}{5} = 1 + \frac{1}{5} \text{ and } 4\frac{5}{6} = 4 + \frac{1}{6} + \frac{1}{6} + \frac{1}{6} + \frac{1}{6} + \frac{1}{6}$$

Every fraction can be built from a combination of unit fractions.

One of the most fundamental concepts of fractions is their ability to be manipulated by multiplication or division. This is possible since $\frac{n}{n} = 1$ for any non-zero integer. As a result, multiplying or dividing by $\frac{n}{n}$ will not alter the original fraction since any number multiplied or divided by 1 doesn't change the value of that number. Fractions of the same value are known as equivalent fractions. For example, $\frac{2}{8}, \frac{25}{100}$, and $\frac{40}{160}$ are equivalent, as they are all equal $\frac{1}{4}$.

Like fractions, or **equivalent fractions**, are the terms used to describe these fractions that are made up of different numbers but represent the same quantity. For example, the given fractions are $^4/_8$ and $^3/_6$. If a pie was cut into 8 pieces and 4 pieces were removed, half of the pie would remain. Also, if a pie was split into 6 pieces and 3 pieces were eaten, half of the pie would also remain. Therefore, both of the fractions represent half of a pie. These two fractions are referred to as like fractions. **Unlike fractions** are fractions that are different and do not represent equal quantities. When working with fractions in mathematical expressions, like fractions should be simplified. Both $^4/_8$ and $^3/_6$ can be simplified into $^1/_2$.

Comparing fractions can be completed through the use of a number line. For example, if $^3/_5$ and $^6/_{10}$ need to be compared, each fraction should be plotted on a number line. To plot $^3/_5$, the area from 0 to 1 should be broken into 5 equal segments, and the fraction represents 3 of them. To plot $^6/_{10}$, the area from 0 to 1 should be broken into 10 equal segments and the fraction represents 6 of them.

It can be seen that $\frac{3}{5} = \frac{6}{10}$

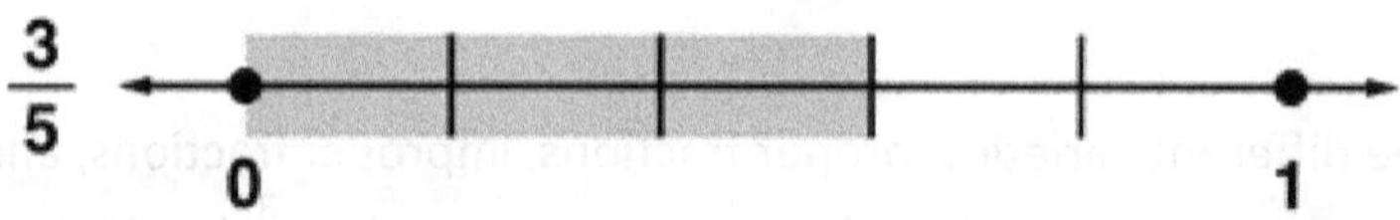

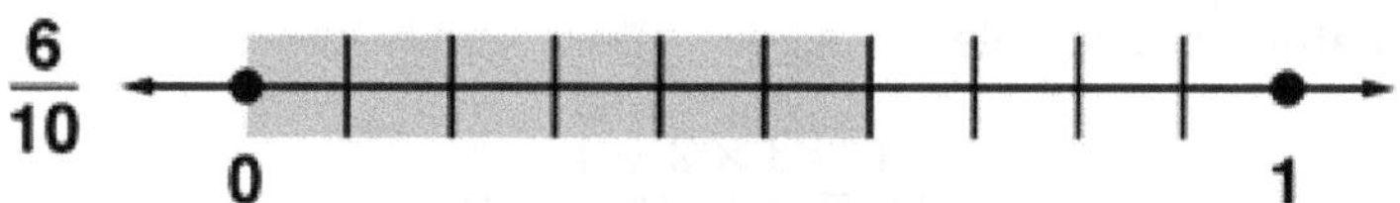

Like fractions are plotted at the same point on a number line. Unit fractions can also be used to compare fractions. For example, if it is known that:

$$\frac{4}{5} > \frac{1}{2}$$

and

$$\frac{1}{2} > \frac{4}{10}$$

then it is also known that:

$$\frac{4}{5} > \frac{4}{10}$$

Also, converting improper fractions to mixed numbers can be helpful in comparing fractions because the whole number portion of the number is more visible.

Adding and subtracting mixed numbers and fractions can be completed by decomposing fractions into a sum of whole numbers and unit fractions. For example, the given problem is:

$$5\frac{3}{7} + 2\frac{1}{7}$$

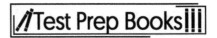

Decomposing into:

$$5 + \frac{1}{7} + \frac{1}{7} + \frac{1}{7} + 2 + \frac{1}{7}$$

This shows that the whole numbers can be added separately from the unit fractions. The answer is:

$$5 + 2 + \frac{1}{7} + \frac{1}{7} + \frac{1}{7} + \frac{1}{7}$$

$$7 + \frac{4}{7}$$

$$7\frac{4}{7}$$

Although many equivalent fractions exist, they are easier to compare and interpret when reduced or simplified. The numerator and denominator of a simple fraction will have no factors in common other than 1. When reducing or simplifying fractions, divide the numerator and denominator by the greatest common factor. A simple strategy is to divide the numerator and denominator by low numbers, like 2, 3, or 5 until arriving at a simple fraction, but the same thing could be achieved by determining the greatest common factor for both the numerator and denominator and dividing each by it. Using the first method is preferable when both the numerator and denominator are even, end in 5, or are obviously a multiple of another number. However, if no numbers seem to work, it will be necessary to factor the numerator and denominator to find the GCF. Let's look at examples:

1) Simplify the fraction $\frac{6}{8}$:

Dividing the numerator and denominator by 2 results in $\frac{3}{4}$, which is a simple fraction.

2) Simplify the fraction $\frac{12}{36}$:

Dividing the numerator and denominator by 2 leaves $\frac{6}{18}$. This isn't a simple fraction, as both the numerator and denominator have factors in common. Diving each by 3 results in $\frac{2}{6}$, but this can be further simplified by dividing by 2 to get $\frac{1}{3}$. This is the simplest fraction, as the numerator is 1. In cases like this, multiple division operations can be avoided by determining the greatest common factor (12, in this case) between the numerator and denominator.

3) Simplify the fraction $\frac{18}{54}$ by dividing by the greatest common factor:

First, determine the factors for the numerator and denominator. The factors of 18 are 1, 2, 3, 6, 9, and 18. The factors of 54 are 1, 2, 3, 6, 9, 18, 27, and 54. Thus, the greatest common factor is 18. Dividing $\frac{18}{54}$ by 18 leaves $\frac{1}{3}$, which is the simplest fraction. This method takes slightly more work, but it definitively arrives at the simplest fraction.

Adding and Subtracting Fractions

Adding and subtracting fractions that have the same denominators involves adding or subtracting the numerators. The denominator will stay the same. Therefore, the decomposition process can be made simpler, and the fractions do not have to be broken into unit fractions.

For example, the given problem is:

$$4\frac{7}{8} - 2\frac{6}{8}$$

The answer is found by adding the answers to both:

$$4 - 2 \text{ and } \frac{7}{8} - \frac{6}{8}$$

$$2 + \frac{1}{8} = 2\frac{1}{8}$$

A common mistake would be to add the denominators so that:

$$\frac{1}{4} + \frac{1}{4} = \frac{1}{8}$$

or to add numerators and denominators so that:

$$\frac{1}{4} + \frac{1}{4} = \frac{2}{8}$$

However, conceptually, it is known that two quarters make a half, so neither one of these are correct.

If two fractions have different denominators, equivalent fractions must be used to add or subtract them. The fractions must be converted into fractions that have common denominators. A **least common denominator** or the product of the two denominators can be used as the common denominator. For example, in the problem $\frac{5}{6} + \frac{2}{3}$, either 6, which is the least common denominator, or 18, which is the product of the denominators, can be used. In order to use 6, $\frac{2}{3}$ must be converted to sixths. A number line can be used to show the equivalent fraction is $\frac{4}{6}$. What happens is that $\frac{2}{3}$ is multiplied by a fractional form of 1 to obtain a denominator of 6. Hence:

$$\frac{2}{3} \times \frac{2}{2} = \frac{4}{6}$$

Therefore, the problem is now:

$$\frac{5}{6} + \frac{4}{6} = \frac{9}{6}$$

which can be simplified into $\frac{3}{2}$. In order to use 18, both fractions must be converted into having 18 as their denominator. $\frac{5}{6}$ would have to be multiplied by $\frac{3}{3}$, and $\frac{2}{3}$ would need to be multiplied by $\frac{6}{6}$.The addition problem would be:

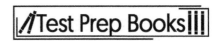

$$\frac{15}{18} + \frac{12}{18} = \frac{27}{18}$$

which reduces into $\frac{3}{2}$.

It is always possible to find a common denominator by multiplying the denominators. However, when the denominators are large numbers, this method is unwieldy, especially if the answer must be provided in its simplest form. Thus, it's beneficial to find the **least common denominator** of the fractions—the least common denominator is incidentally also the **least common multiple**.

Once equivalent fractions have been found with common denominators, simply add or subtract the numerators to arrive at the answer:

1) $\frac{1}{2} + \frac{3}{4} = \frac{2}{4} + \frac{3}{4} = \frac{5}{4}$

2) $\frac{3}{12} + \frac{11}{20} = \frac{15}{60} + \frac{33}{60} = \frac{48}{60} = \frac{4}{5}$

3) $\frac{7}{9} - \frac{4}{15} = \frac{35}{45} - \frac{12}{45} = \frac{23}{45}$

4) $\frac{5}{6} - \frac{7}{18} = \frac{15}{18} - \frac{7}{18} = \frac{8}{18} = \frac{4}{9}$

Multiplying and Dividing Fractions

Of the four basic operations that can be performed on fractions, the one which involves the least amount of work is multiplication. To multiply two fractions, simply multiply the numerators, multiply the denominators, and place the products as a fraction. Whole numbers and mixed numbers can also be expressed as a fraction, as described above, to multiply with a fraction.

Because multiplication is commutative, multiplying a fraction by a whole number is the same as multiplying a whole number by a fraction. The problem involves adding a fraction a specific number of times. The problem $3 \times \frac{1}{4}$ can be translated into adding the unit fraction three times:

$$\frac{1}{4} + \frac{1}{4} + \frac{1}{4} = \frac{3}{4}$$

In the problem $4 \times \frac{2}{5}$, the fraction can be decomposed into $\frac{1}{5} + \frac{1}{5}$ and then added four times to obtain $\frac{8}{5}$. Also, both of these answers can be found by just multiplying the whole number by the numerator of the fraction being multiplied.

The whole numbers can be written in fraction form as:

$$\frac{3}{1} \times \frac{1}{4} = \frac{3}{4}$$

$$\frac{4}{1} \times \frac{2}{5} = \frac{8}{5}$$

Multiplying a fraction times a fraction involves multiplying the numerators together separately and the denominators together separately. For example:

$$\frac{3}{8} \times \frac{2}{3} = \frac{3 \times 2}{8 \times 3} = \frac{6}{24}$$

This can then be reduced to ${}^{1}/_{4}$.

Dividing a fraction by a fraction is actually a multiplication problem. It involves flipping the divisor and then multiplying normally. For example:

$$\frac{22}{5} \div \frac{1}{2} = \frac{22}{5} \times \frac{2}{1} = \frac{44}{5}$$

The same procedure can be implemented for division problems involving fractions and whole numbers. The whole number can be rewritten as a fraction over a denominator of 1, and then division can be completed.

A common denominator approach can also be used in dividing fractions. Considering the same problem:

$$\frac{22}{5} \div \frac{1}{2}$$

a common denominator between the two fractions is 10.

$\frac{22}{5}$ would be rewritten as:

$$\frac{22}{5} \times \frac{2}{2} = \frac{44}{10}$$

and $\frac{1}{2}$ would be rewritten as:

$$\frac{1}{2} \times \frac{5}{5} = \frac{5}{10}$$

Dividing both numbers straight across results in:

$$\frac{44}{10} \div \frac{5}{10} = \frac{{}^{44}/_{5}}{{}^{10}/_{10}} = \frac{{}^{44}/_{5}}{1} = {}^{44}/_{5}$$

Many real-world problems will involve the use of fractions. Key words include actual fraction values, such as *half, quarter, third, fourth,* etc. The best approach to solving word problems involving fractions is to draw a picture or diagram that represents the scenario being discussed, while deciding which type of operation is necessary in order to solve the problem. A phrase such as "one fourth of 60 pounds of coal" creates a scenario in which multiplication should be used, and the mathematical form of the phrase is $\frac{1}{4} \times 60$.

Decimals

The **decimal system** is a way of writing out numbers that uses ten different numerals: 0, 1, 2, 3, 4, 5, 6, 7, 8, and 9. This is also called a "base ten" or "base 10" system. Other bases are also used. For example, computers work with a base of 2. This means they only use the numerals 0 and 1.

The **decimal place** denotes how far to the right of the decimal point a numeral is. The first digit to the right of the decimal point is in the **tenths'** place. The next is the **hundredths'** place. The third is the **thousandths'** place.

So, 3.142 has a 1 in the tenths place, a 4 in the hundredths place, and a 2 in the thousandths place.

The **decimal point** is a period used to separate the **ones'** place from the **tenths'** place when writing out a number as a decimal.

A **decimal number** is a number written out with a decimal point instead of as a fraction, for example, 1.25 instead of $\frac{5}{4}$. Depending on the situation, it may be easier to work with fractions, while other times, it may be easier to work with decimal numbers.

A decimal number is **terminating** if it stops at some point. It is called **repeating** if it never stops but repeats a pattern over and over. It is important to note that every rational number can be written as a terminating decimal or as a repeating decimal.

Addition with Decimals

To add decimal numbers, each number in columns needs to be lined up by the decimal point. For each number being added, the zeros to the right of the last number need to be filled in so that each of the numbers has the same number of places to the right of the decimal. Then, the columns can be added together. Here is an example of 2.45 + 1.3 + 8.891 written in column form:

$$\begin{array}{r} 2.450 \\ 1.300 \\ \underline{+\ 8.891} \end{array}$$

Zeros have been added in the columns so that each number has the same number of places to the right of the decimal.

Added together, the correct answer is 12.641:

$$\begin{array}{r} 2.450 \\ 1.300 \\ \underline{+\ 8.891} \\ 12.641 \end{array}$$

Subtraction with Decimals

Subtracting decimal numbers is the same process as adding decimals. Here is 7.89 – 4.235 written in column form:

$$\begin{array}{r} 7.890 \\ -\underline{4.235} \\ 3.655 \end{array}$$

A zero has been added in the column so that each number has the same number of places to the right of the decimal.

Multiplication with Decimals

Decimals can be multiplied as if there were no decimal points in the problem. For example, 0.5 x 1.25 can be rewritten and multiplied as 5 x 125, which equals 625.

The final answer will have the same number of decimal places as the total number of decimal places in the problem. The first number has one decimal place, and the second number has two decimal places. Therefore, the final answer will contain three decimal places:

$$0.5 \times 1.25 = 0.625$$

Division with Decimals

Dividing a decimal by a whole number entails using long division first by ignoring the decimal point. Then, the decimal point is moved the number of places given in the problem.

For example, $6.8 \div 4$ can be rewritten as $68 \div 4$, which is 17. There is one non-zero integer to the right of the decimal point, so the final solution would have one decimal place to the right of the solution. In this case, the solution is 1.7.

Dividing a decimal by another decimal requires changing the divisor to a whole number by moving its decimal point. The decimal place of the dividend should be moved by the same number of places as the divisor. Then, the problem is the same as dividing a decimal by a whole number.

For example, $5.72 \div 1.1$ has a divisor with one decimal point in the denominator. The expression can be rewritten as $57.2 \div 11$ by moving each number one decimal place to the right to eliminate the decimal. The long division can be completed as $572 \div 11$ with a result of 52. Since there is one non-zero integer to the right of the decimal point in the problem, the final solution is 5.2.

In another example, $8 \div 0.16$ has a divisor with two decimal points in the denominator. The expression can be rewritten as $800 \div 16$ by moving each number two decimal places to the right to eliminate the decimal in the divisor. The long division can be completed with a result of 50.

Percentages

Think of percentages as fractions with a denominator of 100. In fact, **percentage** means "per hundred." Problems often require converting numbers from percentages, fractions, and decimals.

The basic percent equation is the following:

$$\frac{is}{of} = \frac{\%}{100}$$

The placement of numbers in the equation depends on what the question asks.

Example 1
Find 40% of 80.

Basically, the problem is asking, "What is 40% of 80?" The 40% is the percent, and 80 is the number to find the percent "of." The equation is:

$$\frac{x}{80} = \frac{40}{100}$$

Solving the equation by cross-multiplication, the problem becomes 100x = 80(40). Solving for x gives the answer: x = 32.

Example 2
What percent of 90 is 20?

The 20 fills in the "is" portion, while 90 fills in the "of." The question asks for the percent, so that will be x, the unknown. The following equation is set up:

$$\frac{20}{90} = \frac{x}{100}$$

Cross-multiplying yields the equation 90x = 20(100). Solving for x gives the answer of 22.2%.

Example 3
30% of what number is 30?

The following equation uses the clues and numbers in the problem:

$$\frac{30}{x} = \frac{30}{100}$$

Cross-multiplying results in the equation 30(100) = 30x. Solving for x gives the answer x = 100.

<u>Conversions</u>
Decimals and Percentages
Since a percentage is based on "per hundred," decimals and percentages can be converted by multiplying or dividing by 100. Practically speaking, this always involves moving the decimal point two places to the right or left, depending on the conversion. To convert a percentage to a decimal, move the decimal point two places to the left and remove the % sign. To convert a decimal to a percentage, move the decimal point two places to the right and add a "%" sign. Here are some examples:

65% = 0.65
0.33 = 33%
0.215 = 21.5%
99.99% = 0.9999
500% = 5.00
7.55 = 755%

Fractions and Percentages

Remember that a percentage is a number per one hundred. So a percentage can be converted to a fraction by making the number in the percentage the numerator and putting 100 as the denominator:

$$43\% = \frac{43}{100}$$

$$97\% = \frac{97}{100}$$

Note that the percent symbol (%) kind of looks like a 0, a 1, and another 0. So think of a percentage like 54% as 54 over 100.

To convert a fraction to a percent, follow the same logic. If the fraction happens to have 100 in the denominator, you're in luck. Just take the numerator and add a percent symbol:

$$\frac{28}{100} = 28\%$$

Otherwise, divide the numerator by the denominator to get a decimal:

$$\frac{9}{12} = 0.75$$

Then convert the decimal to a percentage:

$$0.75 = 75\%$$

Another option is to make the denominator equal to 100. Be sure to multiply the numerator by the same number as the denominator. For example:

$$\frac{3}{20} \times \frac{5}{5} = \frac{15}{100}$$

$$\frac{15}{100} = 15\%$$

Changing Fractions to Decimals

To change a fraction into a decimal, divide the denominator into the numerator until there are no remainders. There may be repeating decimals, so rounding is often acceptable. A straight line above the repeating portion denotes that the decimal repeats.

Example: Express 4/5 as a decimal.

Set up the division problem.

$$5\overline{)4}$$

5 does not go into 4, so place the decimal and add a zero.

$$5\overline{)4.0}$$

5 goes into 40 eight times. There is no remainder.

```
   0 . 8
 5|4 . 0
 - 4 . 0
       0
```

The solution is 0.8.

Example: Express 33 1/3 as a decimal.

Since the whole portion of the number is known, set it aside to calculate the decimal from the fraction portion.

Set up the division problem.

```
3|1
```

3 does not go into 1, so place the decimal and add zeros. 3 goes into 10 three times.

```
  0 . 3
3|1 . 0
```

This will repeat with a remainder of 1.

```
  0 . 3 3 3
3|1 . 0 0 0
   - 9
     1 0
     - 9
       1 0
```

So, we will place a line over the 3 to denote the repetition. The solution is written $0.\overline{3}$.

Changing Decimals to Fractions

To change decimals to fractions, place the decimal portion of the number—the numerator—over the respective place value—the denominator—then reduce, if possible.

Example: Express 0.25 as a fraction.

This is read as twenty-five hundredths, so put 25 over 100. Then reduce to find the solution.

$$\frac{25}{100} = \frac{1}{4}$$

Example: Express 0.455 as a fraction

This is read as four hundred fifty-five thousandths, so put 455 over 1000. Then reduce to find the solution.

$$\frac{455}{1000} = \frac{91}{200}$$

There are two types of problems that commonly involve percentages. The first is to calculate some percentage of a given quantity, where you convert the percentage to a decimal, and multiply the quantity by that decimal. Secondly, you are given a quantity and told it is a fixed percent of an unknown quantity. In this case, convert to a decimal, then divide the given quantity by that decimal.

Example: What is 30% of 760?

Convert the percent into a useable number. "Of" means to multiply.

$$30\% = 0.30$$

Set up the problem based on the givens, and solve.

$$0.30 \times 760 = 228$$

Example: 8.4 is 20% of what number?

Convert the percent into a useable number.

$$20\% = 0.20$$

The given number is a percent of the answer needed, so divide the given number by this decimal rather than multiplying it.

$$\frac{8.4}{0.20} = 42$$

Factorization

Factors are the numbers multiplied to achieve a product. Thus, every product in a multiplication equation has, at minimum, two factors. Of course, some products will have more than two factors. For the sake of most discussions, assume that factors are positive integers.

To find a number's factors, start with 1 and the number itself. Then divide the number by 2, 3, 4, and so on, seeing if any divisors can divide the number without a remainder, keeping a list of those that do. Stop upon reaching either the number itself or another factor.

Let's find the factors of 45. Start with 1 and 45. Then try to divide 45 by 2, which fails. Now divide 45 by 3. The answer is 15, so 3 and 15 are now factors. Dividing by 4 doesn't work, and dividing by 5 leaves 9. Lastly, dividing 45 by 6, 7, and 8 all don't work. The next integer to try is 9, but this is already known to be a factor, so the factorization is complete. The factors of 45 are 1, 3, 5, 9, 15 and 45.

Prime Factorization

Prime factorization involves an additional step after breaking a number down to its factors: breaking down the factors until they are all prime numbers. A **prime number** is any number that can only be divided by 1 and itself. The prime numbers between 1 and 20 are 2, 3, 5, 7, 11, 13, 17, and 19. As a simple test, numbers that are even or end in 5 are not prime, though there are other numbers that are not prime, but are odd and do not end in 5. For example, 21 is odd and divisible by 1, 3, 7, and 21, so it is not prime.

Let's break 129 down into its prime factors. First, the factors are 3 and 43. Both 3 and 43 are prime numbers, so we're done. But if 43 was not a prime number, then it would also need to be factorized until all of the factors are expressed as prime numbers.

Common Factor

A **common factor** is a factor shared by two numbers. Let's take 45 and 30 and find the common factors:

The factors of 45 are: 1, 3, 5, 9, 15, and 45.
The factors of 30 are: 1, 2, 3, 5, 6, 10, 15, and 30.
Thus, the common factors are 1, 3, 5, and 15.

Greatest Common Factor

The **greatest common factor** is the largest number among the shared, common factors. From the factors of 45 and 30, the common factors are 3, 5, and 15. Therefore, 15 is the greatest common factor, as it's the largest number.

Least Common Multiple

The **least common multiple** is the smallest number that's a multiple of two numbers. Let's try to find the least common multiple of 4 and 9. The multiples of 4 are 4, 8, 12, 16, 20, 24, 28, 32, 36, and so on. For 9, the multiples are 9, 18, 27, 36, 45, 54, etc. Thus, the least common multiple of 4 and 9 is 36 because this is the lowest number where 4 and 9 share multiples.

If two numbers share no factors besides 1 in common, then their least common multiple will be simply their product. If two numbers have common factors, then their least common multiple will be their product divided by their greatest common factor. This can be visualized by the formula:

$$LCM = \frac{x \times y}{GCF}$$

where *x* and *y* are some integers and *LCM* and *GCF* are their least common multiple and greatest common factor, respectively.

Exponents

Exponents are used in mathematics to express a number or variable multiplied by itself a certain number of times. For example, x^3 means *x* is multiplied by itself three times. In this expression, x is called the **base**, and 3 is the **exponent.** Exponents can be used in more complex problems when they contain fractions and negative numbers.

Fractional exponents can be explained by looking first at the inverse of exponents, which are **roots.** Given the expression x^2, the square root can be taken, $\sqrt{x^2}$, cancelling out the 2 and leaving *x* by itself, if *x* is positive. Cancellation occurs because $\sqrt{x}$ can be written with exponents, instead of roots, as $x^{\frac{1}{2}}$. The numerator of 1 is the exponent, and the denominator of 2 is called the **root** (which is why it's referred to as a **square root**). Taking the square root of x^2 is the same as raising it to the $\frac{1}{2}$ power. Written out in mathematical form, it takes the following progression:

$$\sqrt{x^2} = (x^2)^{\frac{1}{2}} = x$$

From properties of exponents, $2 \times \frac{1}{2} = 1$ is the actual exponent of *x*. Another example can be seen with $x^{\frac{4}{7}}$. The variable *x*, raised to four-sevenths, is equal to the seventh root of *x* to the fourth power: $\sqrt[7]{x^4}$. In general:

$$x^{\frac{1}{n}} = \sqrt[n]{x}$$

and

$$x^{\frac{m}{n}} = \sqrt[n]{x^m}$$

Negative exponents also involve fractions. Whereas y^3 can also be rewritten as $\frac{y^3}{1}$, y^{-3} can be rewritten as $\frac{1}{y^3}$. A negative exponent means the exponential expression must be moved to the opposite spot in a fraction to make the exponent positive. If the negative appears in the numerator, it moves to the denominator. If the negative appears in the denominator, it is moved to the numerator. In general, $a^{-n} = \frac{1}{a^n}$, and a^{-n} and a^n are reciprocals.

Take, for example, the following expression:

$$\frac{a^{-4}b^2}{c^{-5}}$$

Since *a* is raised to the negative fourth power, it can be moved to the denominator. Since *c* is raised to the negative fifth power, it can be moved to the numerator. The *b* variable is raised to the positive second power, so it does not move.

The simplified expression is as follows:

$$\frac{b^2c^5}{a^4}$$

In mathematical expressions containing exponents and other operations, the order of operations must be followed. **PEMDAS** states that exponents are calculated after any parenthesis and grouping symbols, but before any multiplication, division, addition, and subtraction.

Roots

The **square root symbol** is expressed as $\sqrt{\ }$ and is commonly known as the **radical**. Taking the root of a number is the inverse operation of multiplying that number by itself some number of times. For example, squaring the number 7 is equal to 7×7, or 49. Finding the square root is the opposite of finding an exponent, as the operation seeks a number that when multiplied by itself, equals the number in the square root symbol.

For example, $\sqrt{36} = 6$ because 6 multiplied by 6 equals 36. Note, the square root of 36 is also -6 since $-6 \times -6 = 36$. This can be indicated using a plus/minus symbol like this: ±6. However, square roots are often just expressed as a positive number for simplicity, with it being understood that the true value can be either positive or negative.

Perfect squares are numbers with whole number square roots. The list of perfect squares begins with 0, 1, 4, 9, 16, 25, 36, 49, 64, 81, and 100.

Determining the square root of imperfect squares requires a calculator to reach an exact figure. It's possible, however, to approximate the answer by finding the two perfect squares that the number fits between. For example, the square root of 40 is between 6 and 7 since the squares of those numbers are 36 and 49, respectively.

Square roots are the most common root operation. If the radical doesn't have a number to the upper left of the symbol $\sqrt{}$, then it's a square root. Sometimes a radical includes a number in the upper left, like $\sqrt[3]{27}$, as in the other common root type—the cube root. Calculating complicated roots, like the cube root, often requires the use of a calculator.

Scientific Notation

Scientific Notation is used to represent numbers that are either very small or very large. For example, the distance to the Sun is approximately 150,000,000,000 meters. Instead of writing this number with so many zeros, it can be written in scientific notation as 1.5×10^{11}meters. The same is true for very small numbers, but the exponent becomes negative. If the mass of a human cell is 0.000000000001 kilograms, that measurement can be easily represented by 1.0×10^{-12} kilograms. In both situations, scientific notation makes the measurement easier to read and understand. Each number is translated to an expression with one digit in the tens place multiplied by an expression corresponding to the zeros.

When two measurements are given and both involve scientific notation, it is important to know how these interact with each other:

- In addition and subtraction, the exponent on the ten must be the same before any operations are performed on the numbers. For example, $(1.3 \times 10^4) + (3.0 \times 10^3)$ cannot be added until one of the exponents on the ten is changed. The 3.0×10^3 can be changed to 0.3×10^4, then the 1.3 and 0.3 can be added. The answer comes out to be 1.6×10^4.
- For multiplication, the first numbers can be multiplied and then the exponents on the tens can be added. Once an answer is formed, it may have to be converted into scientific notation again depending on the change that occurred.
- The following is an example of multiplication with scientific notation:

$$(4.5 \times 10^3) \times (3.0 \times 10^{-5}) = 13.5 \times 10^{-2}$$

- Since this answer is not in scientific notation, the decimal is moved over to the left one unit, and 1 is added to the ten's exponent. This results in the final answer: 1.35×10^{-1}.

- For division, the first numbers are divided, and the exponents on the tens are subtracted. Again, the answer may need to be converted into scientific notation form, depending on the type of changes that occurred during the problem.

- **Order of magnitude** relates to scientific notation and is the total count of powers of 10 in a number. For example, there are 6 orders of magnitude in 1,000,000. If a number is raised by an order of magnitude, it is multiplied by 10. Order of magnitude can be helpful in estimating results using very large or small numbers. An answer should make sense in terms of its order of magnitude. For example, if area is calculated using two dimensions with 6 orders of magnitude, because area involves multiplication, the answer should have around 12 orders of magnitude. Also, answers can be estimated by rounding to the largest place value in each number. For example, $5{,}493{,}302 \times 2{,}523{,}100$ can be estimated by $5 \times 3 = 15$ with 6 orders of magnitude.

Estimation

Estimation is finding a value that is close to a solution, but is not the exact answer. For example, if there are values in the thousands to be multiplied, then each value can be estimated to the nearest thousand and the calculation performed. This value provides an approximate solution that can be determined very quickly.

Rounding is the process of either bumping a number up or down, based on a specified place value. First, the place value is specified. Then, the digit to its right is looked at. For example, if rounding to the nearest hundreds place, the digit in the tens place is used. If it is a 0, 1, 2, 3, or 4, the digit being rounded to is left alone. If it is a 5, 6, 7, 8 or 9, the digit being rounded to is increased by one. All other digits before the decimal point are then changed to zeros, and the digits in decimal places are dropped. If a decimal place is being rounded to, all subsequent digits are just dropped. For example, if 845,231.45 was to be rounded to the nearest thousands place, the answer would be 845,000. The 5 would remain the same due to the 2 in the hundreds place. Also, if 4.567 was to be rounded to the nearest tenths place, the answer would be 4.6. The 5 increased to 6 due to the 6 in the hundredths place, and the rest of the decimal is dropped.

Sometimes when performing operations such as multiplying numbers, the result can be estimated by rounding. For example, to estimate the value of 11.2×2.01, each number can be rounded to the nearest integer. This will yield a result of 22.

Rounding numbers helps with estimation because it changes the given number to a simpler, although less accurate, number than the exact given number. Rounding allows for easier calculations, which estimate the results of using the exact given number. The accuracy of the estimate and ease of use depends on the place value to which the number is rounded. Rounding numbers consists of:

- determining what place value the number is being rounded to
- examining the digit to the right of the desired place value to decide whether to round up or keep the digit, and
- replacing all digits to the right of the desired place value with zeros.

To round 746,311 to the nearest ten thousand, the digit in the ten thousands place should be located first. In this case, this digit is 4 (7$\underline{4}$6,311). Then, the digit to its right is examined. If this digit is 5 or greater, the number will be rounded up by increasing the digit in the desired place by one. If the digit to the right of the place value being rounded is 4 or less, the number will be kept the same. For the given example, the digit being examined is a 6, which means that the number will be rounded up by increasing the digit to the left by one. Therefore, the digit 4 is changed to a 5. Finally, to write the rounded number, any digits to the left of the place value being rounded remain the same and any to its right are replaced with zeros. For the given example, rounding 746,311 to the nearest ten thousand will produce 750,000. To round 746,311 to the nearest hundred, the digit to the right of the three in the hundreds place is examined to determine whether to round up or keep the same number. In this case, that digit is a 1, so the number will be kept the same and any digits to its right will be replaced with zeros. The resulting rounded number is 746,300.

Rounding place values to the right of the decimal follows the same procedure, but digits being replaced by zeros can simply be dropped. To round 3.752891 to the nearest thousandth, the desired place value is located (3.75<u>2</u>891) and the digit to the right is examined. In this case, the digit 8 indicates that the number will be rounded up, and the 2 in the thousandths place will increase to a 3. Rounding up and replacing the digits to the right of the thousandths place produces 3.753000 which is equivalent to 3.753. Therefore, the zeros are not necessary and the rounded number should be written as 3.753.

When rounding up, if the digit to be increased is a 9, the digit to its left is increased by 1 and the digit in the desired place value is changed to a zero. For example, the number 1,598 rounded to the nearest ten is 1,600. Another example shows the number 43.72961 rounded to the nearest thousandth is 43.730 or 43.73.

Mental math should always be considered as problems are worked through, and the ability to work through problems in one's head helps save time. If a problem is simple enough, such as $15 + 3 = 18$, it should be completed mentally. The ability to do this will increase once addition and subtraction in higher place values are grasped. Also, mental math is important in multiplication and division. The times tables multiplying all numbers from 1 to 12 should be memorized. This will allow for division within those numbers to be memorized as well. For example, we should know easily that $121 \div 11 = 11$ because it should be memorized that $11 \times 11 = 121$.

Here is the multiplication table to be memorized:

x	1	2	3	4	5	6	7	8	9	10	11	12	13	14	15
1	1	2	3	4	5	6	7	8	9	10	11	12	13	14	15
2	2	4	6	8	10	12	14	16	18	20	22	24	26	28	30
3	3	6	9	12	15	18	21	24	27	30	33	36	39	42	45
4	4	8	12	16	20	24	28	32	36	40	44	48	52	56	60
5	5	10	15	20	25	30	35	40	45	50	55	60	65	70	75
6	6	12	18	24	30	36	42	48	54	60	66	72	78	84	90
7	7	14	21	28	35	42	49	56	63	70	77	84	91	98	105
8	8	16	24	32	40	48	56	64	72	80	88	96	104	112	120
9	9	18	27	36	45	54	63	72	81	90	99	108	117	126	135
10	10	20	30	40	50	60	70	80	90	100	110	120	130	140	150
11	11	22	33	44	55	66	77	88	99	110	121	132	143	154	165
12	12	24	36	48	60	72	84	96	108	120	132	144	156	168	180
13	13	26	39	52	65	78	91	104	117	130	143	156	169	182	195
14	14	28	42	56	70	84	98	112	126	140	154	168	182	196	210
15	15	30	45	60	75	90	105	120	135	150	165	180	195	210	225

The values in gray along the diagonal of the table consist of **perfect squares**. A perfect square is a number that represents a product of two equal integers.

Sequences and Series

Patterns within a sequence can come in 2 distinct forms: the items (shapes, numbers, etc.) either repeat in a constant order, or the items change from one step to another in some consistent way. The **core** is the smallest unit, or number of items, that repeats in a repeating pattern. For example, the pattern

oo▲oo▲o…

has a core that is oo▲. Knowing only the core, the pattern can be extended. Knowing the number of steps in the core allows the identification of an item in each step without drawing/writing the entire pattern out. For example, suppose the tenth item in the previous pattern must be determined. Because the core consists of three items (oo▲), the core repeats in multiples of 3. In other words, steps 3, 6, 9, 12, etc. will be ▲ completing the core with the core starting over on the next step. For the above example, the 9th step will be ▲ and the 10th will be o.

The most common patterns in which each item changes from one step to the next are arithmetic and geometric sequences. An **arithmetic sequence** is one in which the items increase or decrease by a constant difference. In other words, the same thing is added or subtracted to each item or step to produce the next. To determine if a sequence is arithmetic, determine what must be added or subtracted to step one to produce step two. Then, check if the same thing is added/subtracted to step two to produce step three. The same thing must be added/subtracted to step three to produce step four, and so on. Consider the pattern 13, 10, 7, 4 . . . To get from step one (13) to step two (10) by adding or subtracting requires subtracting by 3. The next step is checking if subtracting 3 from step two (10) will produce step three (7), and subtracting 3 from step three (7) will produce step four (4). In this case, the pattern holds true. Therefore, this is an arithmetic sequence in which each step is produced by subtracting 3 from the previous step. To extend the sequence, 3 is subtracted from the last step to produce the next. The next three numbers in the sequence are 1, -2, -5.

A **geometric sequence** is one in which each step is produced by multiplying or dividing the previous step by the same number. To determine if a sequence is geometric, decide what step one must be multiplied or divided by to produce step two. Then check if multiplying or dividing step two by the same number produces step three, and so on. Consider the pattern 2, 8, 32, 128 . . . To get from step one (2) to step two (8) requires multiplication by 4. The next step determines if multiplying step two (8) by 4 produces step three (32), and multiplying step three (32) by 4 produces step four (128). In this case, the pattern holds true. Therefore, this is a geometric sequence in which each step is produced by multiplying the previous step by 4. To extend the sequence, the last step is multiplied by 4 and repeated. The next three numbers in the sequence are 512, 2,048, and 8,192.

Although arithmetic and geometric sequences typically use numbers, these sequences can also be represented by shapes. For example, an arithmetic sequence could consist of shapes with three sides, four sides, and five sides (add one side to the previous step to produce the next). A geometric sequence could consist of eight blocks, four blocks, and two blocks (each step is produced by dividing the number of blocks in the previous step by 2).

Frequencies

In mathematics, **frequencies** refer to how often an event occurs or the number of times a particular quantity appears in a given series. To find the number of times a specific value appears, frequency tables are used to record the occurrences, which can then be summed. To construct a frequency table, one simply inputs the values into a tabular format with a column denoting each value, typically in ascending order, with a second column to tally up the number of occurrences for each value, and a third column to give a numerical frequency based on the number of tallies.

A **frequency distribution** communicates the number of outcomes of a given value or number in a data set. When displayed as a bar graph or histogram, it can visually indicate the spread and distribution of the data. A histogram resembling a bell curve approximates a normal distribution.

A frequency distribution can also be displayed as a **stem-and-leaf plot**, which arranges data in numerical order and displays values similar to a tally chart with the stem being a range within the set and the leaf indicating the exact value. (Ex. stems are whole numbers and leaves are tenths.)

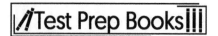

	Movie Ratings							
4	7							
5	2	6	9					
6	1	4	6	8	8			
7	0	3	5	9				
8	1	3	5	6	8	8	9	
9	0	0	1	3	4	6	6	9
Key	6 | 1 represents 61							

Stem	Leaf						
2	0	2	3	6	8	8	9
3	2	6	7	7			
4	7	9					
5	4	6	9				

This plot provides more detail about individual data points and allows for easy identification of the median, as well as any repeated values in the set.

Data that isn't described using numbers is known as **categorical data.** For example, age is numerical data but hair color is categorical data. Categorical data can also be summarized using two-way frequency tables. A **two-way frequency table** counts the relationship between two sets of categorical data. There are rows and columns for each category, and each cell represents frequency information that shows the actual data count between each combination. For example, below is a two-way frequency table showing the gender and breed of cats in an animal shelter:

	Domestic Shorthair	**Persian**	**Domestic Longhair**	**Total**
Male	12	2	7	21
Female	8	4	5	17
Total	20	6	12	38

Entries in the middle of the table are known as the **joint frequencies**. For example, the number of females that are Persians is 4, which is a joint frequency. The totals are the **marginal frequencies**. For example, the total number of males is 21, which is a marginal frequency. If the frequencies are changed into percentages based on totals, the table is known as a **two-way relative frequency table**. Percentages can be calculated using the table total, the row totals, or the column totals. Two-way frequency tables can help in making conclusions about the data.

Solving Word Problems

Word problems, or story problems, are math problems that have a real-world context. In word problems, multiple quantities are often provided with a request to find some kind of relation between them. This often will mean that one variable (the dependent variable whose value needs to be found) can be written as a function of another variable (the independent variable whose value can be figured from the given information). The usual procedure for solving these problems is to start by giving each quantity in the problem a variable, and then figuring the relationship between these variables.

For example, suppose a car gets 25 miles per gallon. How far will the car travel if it uses 2.4 gallons of fuel? In this case, *y* would be the distance the car has traveled in miles, and *x* would be the amount of fuel burned in gallons (2.4). Then the relationship between these variables can be written as an algebraic equation, $y = 25x$. In this case, the equation is:

$$y = 25 \times 2.4 = 60$$

so the car has traveled 60 miles.

Translating Verbal Relationships into Algebraic Equations or Expressions

When attempting to solve a math problem, it's important to apply the correct algorithm. It is much more difficult to determine what algorithm is necessary when solving word problems, because the necessary operations and equations are typically not provided. In these instances, the test taker must translate the words in the problem into true mathematical statements that can be solved.

The following are examples:

Symbol	Phrase
+	Added to; increased by; sum of; more than
−	Decreased by; difference between; less than; take away
×	Multiplied by; 3(4,5...) times as large; product of
÷	Divided by; quotient of; half (third, etc.) of
=	Is; the same as; results in; as much as; equal to
x, t, n, etc.	A number; unknown quantity; value of; variable

As mentioned, addition and subtraction are **inverse operations**. Adding a number and then subtracting the same number will cancel each other out, resulting in the original number, and vice versa. For example:

$$8 + 7 - 7 = 8 \text{ and } 137 - 100 + 100 = 137$$

Similarly, multiplication and division are inverse operations. Therefore, multiplying by a number and then dividing by the same number results in the original number, and vice versa. For example:

$$8 \times 2 \div 2 = 8 \text{ and } 12 \div 4 \times 4 = 12$$

Inverse operations are used to work backwards to solve problems. In the case that 7 and a number add to 18, the inverse operation of subtraction is used to find the unknown value ($18 - 7 = 11$). If a school's entire 4th grade was divided evenly into 3 classes each with 22 students, the inverse operation of multiplication is used to determine the total students in the grade ($22 \times 3 = 66$). Additional scenarios involving inverse operations are included in the tables below.

There are a variety of real-world situations in which one or more of the operators is used to solve a problem. The tables below display the most common scenarios.

Addition & Subtraction

	Unknown Result	**Unknown Change**	**Unknown Start**
Adding to	5 students were in class. 4 more students arrived. How many students are in class? $5+4=?$	8 students were in class. More students arrived late. There are now 18 students in class. How many students arrived late? $8+?=18$ Solved by inverse operations $18-8=?$	Some students were in class early. 11 more students arrived. There are now 17 students in class. How many students were in class early? $?+11=17$ Solved by inverse operations $17-11=?$
Taking from	15 students were in class. 5 students left class. How many students are in class now? $15-5=?$	12 students were in class. Some students left class. There are now 8 students in class. How many students left class? $12-?=8$ Solved by inverse operations $8+?=12 \rightarrow 12-8=?$	Some students were in class. 3 students left class. Then there were 13 students in class. How many students were in class before? $?-3=13$ Solved by inverse operations $13+3=?$

	Unknown Total	**Unknown Addends (Both)**	**Unknown Addends (One)**
Putting together/ taking apart	The homework assignment is 10 addition problems and 8 subtraction problems. How many problems are in the homework assignment? $10+8=?$	Bobby has $9. How much can Bobby spend on candy and how much can Bobby spend on toys? $9=?+?$	Bobby has 12 pairs of pants. 5 pairs of pants are shorts, and the rest are long. How many pairs of long pants does he have? $12=5+?$ Solved by inverse operations $12-5=?$

	Unknown Difference	Unknown Larger Value	Unknown Smaller Value
Comparing	Bobby has 5 toys. Tommy has 8 toys. How many more toys does Tommy have than Bobby? $5+?=8$ Solved by inverse operations $8-5=?$ Bobby has $6. Tommy has $10. How many fewer dollars does Bobby have than Tommy? $10-6=?$	Tommy has 2 more toys than Bobby. Bobby has 4 toys. How many toys does Tommy have? $2+4=?$ Bobby has 3 fewer dollars than Tommy. Bobby has $8. How many dollars does Tommy have? $?-3=8$ Solved by inverse operations $8+3=?$	Tommy has 6 more toys than Bobby. Tommy has 10 toys. How many toys does Bobby have? $?+6=10$ Solved by inverse operations $10-6=?$ Bobby has $5 less than Tommy. Tommy has $9. How many dollars does Bobby have? $9-5=?$

Multiplication and Division

	Unknown Product	Unknown Group Size	Unknown Number of Groups
Equal groups	There are 5 students, and each student has 4 pieces of candy. How many pieces of candy are there in all? $5\times 4=?$	14 pieces of candy are shared equally by 7 students. How many pieces of candy does each student have? $7\times ?=14$ Solved by inverse operations $14\div 7=?$	If 18 pieces of candy are to be given out 3 to each student, how many students will get candy? $?\times 3=18$ Solved by inverse operations $18\div 3=?$

	Unknown Product	Unknown Factor	Unknown Factor
Arrays	There are 5 rows of students with 3 students in each row. How many students are there? $5\times 3=?$	If 16 students are arranged into 4 equal rows, how many students will be in each row? $4\times ?=16$ Solved by inverse operations $16\div 4=?$	If 24 students are arranged into an array with 6 columns, how many rows are there? $?\times 6=24$ Solved by inverse operations $24\div 6=?$

	Larger Unknown	Smaller Unknown	Multiplier Unknown
Comparing	A small popcorn costs \$1.50. A large popcorn costs 3 times as much as a small popcorn. How much does a large popcorn cost? $1.50 \times 3 = ?$	A large soda costs \$6 and that is 2 times as much as a small soda costs. How much does a small soda cost? $2 \times ? = 6$ Solved by inverse operations $6 \div 2 = ?$	A large pretzel costs \$3 and a small pretzel costs \$2. How many times as much does the large pretzel cost as the small pretzel? $? \times 2 = 3$ Solved by inverse operations $3 \div 2 = ?$

Modeling and Solving Word Problems

Word problems can appear daunting, but don't let the wording intimidate you. No matter the scenario or specifics, the key to answering them is to translate the words into a math problem. Always keep in mind what the question is asking and what operations could lead to that answer.

Some word problems require more than just one simple equation to be written and solved. Consider the following situations and the linear equations used to model them.

Suppose Margaret is 2 miles to the east of John at noon. Margaret walks to the east at 3 miles per hour. How far apart will they be at 3 p.m.? To solve this, *x* would represent the time in hours past noon, and *y* would represent the distance between Margaret and John. Now, noon corresponds to the equation where *x* is 0, so the *y* intercept is going to be 2. It's also known that the slope will be the rate at which the distance is changing, which is 3 miles per hour. This means that the slope will be 3 (be careful at this point: if units were used, other than miles and hours, for *x* and *y* variables, a conversion of the given information to the appropriate units would be required first). The simplest way to write an equation given the *y*-intercept, and the slope is the Slope-Intercept form, is $y = mx + b$. Recall that *m* here is the slope and *b* is the *y* intercept. So, $m = 3$ and $b = 2$. Therefore, the equation will be $y = 3x + 2$. The word problem asks how far to the east Margaret will be from John at 3 p.m., which means when *x* is 3. So, substitute $x = 3$ into this equation to obtain:

$$y = 3 \cdot 3 + 2 = 9 + 2 = 11$$

Therefore, she will be 11 miles to the east of him at 3 p.m.

For another example, suppose that a box with 4 cans in it weighs 6 lbs., while a box with 8 cans in it weighs 12 lbs. Find out how much a single can weighs. To do this, let *x* denote the number of cans in the box, and *y* denote the weight of the box with the cans in lbs. This line touches two pairs: $(4, 6)$ and (8, 12). A formula for this relation could be written using the two-point form, with:

$$x_1 = 4, y_1 = 6, x_2 = 8, y_2 = 12$$

This would yield:

$$\frac{y-6}{x-4}=\frac{12-6}{8-4}$$

$$\frac{y-6}{x-4}=\frac{6}{4}=\frac{3}{2}$$

However, only the slope is needed to solve this problem, since the slope will be the weight of a single can. From the computation, the slope is $\frac{3}{2}$. Therefore, each can weighs $\frac{3}{2}$ lb.

The Problem-Solving Process and Determining If Enough Information Is Provided to Solve a Problem

Overall, the problem-solving process in mathematics involves a step-by-step procedure that one must follow when deciding what approach to take. First, one must understand the problem by deciding what is being sought, then if enough information is given, and what units are necessary in the solution. This is a crucial, but sometimes difficult step. It involves carefully reading the entire problem, identifying (perhaps even underlining) the facts or information that *is* known, and then deciphering the question words to determine what the problem is asking. In this way, math problems require students to be detectives, evaluating the "clues" or facts given in the problem, deciding what the problem is looking for, and evaluating whether sufficient information or "clues" are presented in the problem to solve the posed question.

In general, when solving word problems (also called story problems), it's important to understand what is being asked and to properly set up the initial equation. Always read the entire problem through, and then separate out what information is given in the statement. Decide what you are being asked to find and label each quantity with a variable or constant. Then write an equation to determine the unknown variable. Remember to label answers; sometimes knowing what the answers' units should can help eliminate other possible solutions.

When trying to solve any word problem, look for a series of key words indicating addition, subtraction, multiplication, or division to help you determine how to set up the problem:

Addition: *add, altogether, together, plus, increased by, more than, in all, sum,* and *total*

Subtraction: *minus, less than, difference, decreased by, fewer than, remain,* and *take away*

Multiplication: *times, twice, of, double,* and *triple*

Division: *divided by, cut up, half, quotient of, split,* and *shared equally*

If a question asks to give words to a mathematical expression and says "equals," then an = sign must be included in the answer. Similarly, "less than or equal to" is expressed by the inequality symbol ≤, and "greater than or equal" to is expressed as ≥. Furthermore, "less than" is represented by <, and "greater than" is expressed by >.

These strategies are applicable to other question types. For example, calculating salary after deductions, balancing a checkbook, and calculating a dinner bill are common word problems similar to business planning. Just remember to use the correct operations. When a balance is increased, use addition.

When a balance is decreased, use subtraction. Common sense and organization are your greatest assets when answering word problems.

For example, suppose the following word problem is encountered:

Walter's Coffee Shop sells a variety of drinks and breakfast treats.

Price List	
Hot Coffee	$2.00
Slow-Drip Iced Coffee	$3.00
Latte	$4.00
Muffin	$2.00
Crepe	$4.00
Egg Sandwich	$5.00

Costs	
Hot Coffee	$0.25
Slow-Drip Iced Coffee	$0.75
Latte	$1.00
Muffin	$1.00
Crepe	$2.00
Egg Sandwich	$3.00

Walter's utilities, rent, and labor costs him $500 per day. Today, Walter sold 200 hot coffees, 100 slow-drip iced coffees, 50 lattes, 75 muffins, 45 crepes, and 60 egg sandwiches. What was Walter's total profit today?

First, it is necessary to establish what is known (the "facts"), what one wants to know, (the question), how to determine the answer (the process), and if there is enough information to solve (sufficient "clues"). The problem clearly asks: "what was Walter's total profit today," so to accurately answer this type of question, the total cost of making his drinks and treats must be calculated, then the total revenue he earned from selling those products must be determined. After arriving at these two totals, the profit is measured found by deducting the total cost from the total revenue.

Now that the question and steps are identified, the provided facts are evaluated. Walter's costs for today:

Item	**Quantity**	**Cost Per Unit**	**Total Cost**
Hot Coffee	200	$0.25	$50
Slow-Drip Iced Coffee	100	$0.75	$75
Latte	50	$1.00	$50
Muffin	75	$1.00	$75
Crepe	45	$2.00	$90
Egg Sandwich	60	$3.00	$180
Utilities, rent, and labor			$500
Total Costs			$1,020

Walter's revenue for today:

Item	Quantity	Revenue Per Unit	Total Revenue
Hot Coffee	200	$2.00	$400
Slow-Drip Iced Coffee	100	$3.00	$300
Latte	50	$4.00	$200
Muffin	75	$2.00	$150
Crepe	45	$4.00	$180
Egg Sandwich	60	$5.00	$300
Total Revenue			$1,530

Walter's Profit = *Revenue* – *Costs* = $1,530 – $1,020 = $510

In this case, enough information was given in the problem to adequately solve it. If, however, the number of sandwiches and drinks or Walter's cost per unit sold were not provided, insufficient information would prevent one from arriving at the answer.

Alternative Methods for Solving Mathematical Problems

When solving a math problem, once the question is identified and the clues are evaluated, the plan of action must be determined. In some cases, there might be many options. Therefore, one should begin with one approach and if the strategy does not fit, he or she should move on to another. In some cases, a combination of approaches can be used. A beginning estimate is always useful for comparison once a solution is found. The answer must be reasonable and must fulfill all requirements of the problem.

Just as there are different types of learners (visual, kinesthetic, etc.), so too are there particular problem-solving approaches that different students prefer or grasp more easily than others. Skilled mathematicians are versed in multiple methods to tackle various problems, with each method bolstering their toolbox with a strategy that can be employed for ease and efficiency when encountering math work.

Instead of focusing on the "right" way to solve a problem, students strive to learn multiple methods and understand the pros, cons, and appropriate applications for each method. For example, when trying to find the zeros in a binomial expression, one might be able to factor the expression, complete the square, use the quadratic equation, or make a rough sketch of the graph and identify the x-intercepts. In some cases, one method may not be possible and another may be "easiest," but by learning the various strategies, students become critical thinkers and select the method they deem most appropriate.

The following two examples demonstrate how different methods can be used for the same problem:

Example:

A store is having a spring sale, where everything is 70% off. You have $45.00 to spend. A jacket is regularly priced at $80.00. Do you have enough to buy the jacket and a pair of gloves, regularly priced at $20.00?

There are two ways to approach this.

Method 1:

Set up the equations to find the sale prices: the original price minus the amount discounted.
$80.00 - ($80.00 (0.70)) = sale cost of the jacket.
$20.00 – ($20.00 (0.70)) = sale cost of the gloves.
Solve for the sale cost.
$24.00 = sale cost of the jacket.
$6.00 = sale cost of the gloves.
Determine if you have enough money for both.
$24.00 + $6.00 = total sale cost.
$30.00 is less than $45.00, so you can afford to purchase both.

Method 2:

Determine the percent of the original price that you will pay.
100% – 70% = 30%
Set up the equations to find the sale prices.
$80.00 (0.30) = cost of the jacket.
$20.00 (0.30) = cost of the gloves.
Solve.
$24.00 = cost of the jacket.
$6.00 = cost of the gloves.
Determine if you have enough money for both.
$24.00 + $6.00 = total sale cost.
$30.00 is less than $45.00, so you can afford to purchase both.

Here's another example:

Mary and Dottie team up to mow neighborhood lawns. If Mary mows 2 lawns per hour and the two of them can mow 17.5 lawns in 5 hours, how many lawns does Dottie mow per hour?

Given rate for Mary.

$$Mary = \frac{2\ lawns}{1\ hour}$$

Unknown rate of D for Dottie.

$$Dottie = \frac{D\ lawns}{1\ hour}$$

Given rate for both.

$$Total\ mowed\ together = \frac{17.5\ lawns}{5\ hours}$$

Set up the equation for what is being asked.

$$Mary + Dottie = total\ together.$$

Fill in the givens.

$$2 + D = \frac{17.5}{5}$$

Divide.

$$2 + D = 3.5$$

Subtract 2 from both sides to isolate the variable.

$$2 - 2 + D = 3.5 - 2$$

Solve and label Dottie's mowing rate.

$$D = 1.5 \textit{ lawns per hour}$$

Algebra

Algebraic Expressions and Equations

An **algebraic expression** is a statement about an unknown quantity expressed in mathematical symbols. A **variable** is used to represent the unknown quantity, usually denoted by a letter. An equation is a statement in which two expressions (at least one containing a variable) are equal to each other. An algebraic expression can be thought of as a mathematical phrase and an equation can be thought of as a mathematical sentence.

Algebraic expressions and equations both contain numbers, variables, and mathematical operations. The following are examples of algebraic expressions:

$$5x + 3, 7xy - 8(x^2 + y)$$

and

$$\sqrt{a^2 + b^2}$$

An expression can be simplified or evaluated for given values of variables. The following are examples of equations:

$$2x + 3 = 7$$

$$a^2 + b^2 = c^2$$

$$2x + 5 = 3x - 2$$

An equation contains two sides separated by an equal sign. Equations can be solved to determine the value(s) of the variable for which the statement is true.

Parts of Expressions

Algebraic expressions consist of variables, numbers, and operations. A **term** of an expression is any combination of numbers and/or variables, and terms are separated by addition and subtraction. For example, the expression:

$$5x^2 - 3xy + 4 - 2$$

consists of 4 terms: $5x^2$, -3*xy*, 4*y*, and -2. Note that each term includes its given sign (+ or −). The **variable** part of a term is a letter that represents an unknown quantity. The **coefficient** of a term is the number by which the variable is multiplied. For the term 4*y*, the variable is *y* and the coefficient is 4. Terms are identified by the power (or exponent) of its variable.

A number without a variable is referred to as a **constant**. If the variable is to the first power (x^1 or simply *x*), it is referred to as a linear term. A term with a variable to the second power (x^2) is quadratic and a term to the third power (x^3) is cubic. Consider the expression:

$$x^3 + 3x - 1$$

The constant is -1. The linear term is 3*x*. There is no quadratic term. The cubic term is x^3.

An algebraic expression can also be classified by how many terms exist in the expression. Any like terms should be combined before classifying. A **monomial** is an expression consisting of only one term. Examples of monomials are: 17, 2*x*, and $-5ab^2$. A **binomial** is an expression consisting of two terms separated by addition or subtraction. Examples include:

$$2x - 4 \text{ and } -3y^2 + 2y$$

A **trinomial** consists of 3 terms. For example, $5x^2 - 2x + 1$ is a trinomial.

Adding and Subtracting Linear Algebraic Expressions

An algebraic expression is simplified by combining like terms. As mentioned, term is a number, variable, or product of a number, and variables separated by addition and subtraction. For the algebraic expression:

$$3x^2 - 4x + 5 - 5x^2 + x - 3$$

the terms are $3x^2$, -4*x*, 5, -$5x^2$, *x*, and -3. Like terms have the same variables raised to the same powers (exponents). The like terms for the previous example are $3x^2$ and -$5x^2$, -4x and x, 5 and -3. To combine like terms, the coefficients (numerical factor of the term including sign) are added and the variables and their powers are kept the same. Note that if a coefficient is not written, it is an implied coefficient of 1 ($x = 1x$). The previous example will simplify to:

$$-2x^2 - 3x + 2$$

When adding or subtracting algebraic expressions, each expression is written in parenthesis. The negative sign is distributed when necessary, and like terms are combined. Consider the following: add

$$2a + 5b - 2 \text{ to } a - 2b + 8c - 4$$

The sum is set as follows:

$$(a - 2b + 8c - 4) + (2a + 5b - 2)$$

In front of each set of parentheses is an implied positive one, which, when distributed, does not change any of the terms. Therefore, the parentheses are dropped and like terms are combined:

$$a - 2b + 8c - 4 + 2a + 5b - 2$$

$$3a + 3b + 8c - 6$$

Consider the following problem: Subtract $2a + 5b - 2$ from $a - 2b + 8c - 4$. The difference is set as follows:

$$(a - 2b + 8c - 4) - (2a + 5b - 2)$$

The implied one in front of the first set of parentheses will not change those four terms. However, distributing the implied -1 in front of the second set of parentheses will change the sign of each of those three terms:

$$a - 2b + 8c - 4 - 2a - 5b + 2$$

Combining like terms yields the simplified expression:

$$-a - 7b + 8c - 2$$

Distributive Property

The distributive property states that multiplying a sum (or difference) by a number produces the same result as multiplying each value in the sum (or difference) by the number and adding (or subtracting) the products. Using mathematical symbols, the distributive property states:

$$a(b + c) = ab + ac$$

The expression $4(3 + 2)$ is simplified using the order of operations. Simplifying inside the parenthesis first produces 4×5, which equals 20.

The expression $4(3 + 2)$ can also be simplified using the distributive property:

$$4(3 + 2)$$

$$4 \times 3 + 4 \times 2$$

$$12 + 8 = 20$$

Consider the following example: $4(3x - 2)$. The expression cannot be simplified inside the parenthesis because 3*x* and -2 are not like terms, and therefore cannot be combined. However, the expression can be simplified by using the distributive property and multiplying each term inside of the parenthesis by the term outside of the parenthesis: $12x - 8$. The resulting equivalent expression contains no like terms, so it cannot be further simplified.

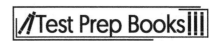

Consider the expression:

$$(3x + 2y + 1) - (5x - 3) + 2(3y + 4)$$

Again, there are no like terms, but the distributive property is used to simplify the expression. Note there is an implied one in front of the first set of parentheses and an implied -1 in front of the second set of parentheses. Distributing the one, -1, and 2 produces:

$$1(3x) + 1(2y) + 1(1) - 1(5x) - 1(-3) + 2(3y) + 2(4)$$

$$3x + 2y + 1 - 5x + 3 + 6y + 8$$

This expression contains like terms that are combined to produce the simplified expression:

$$-2x + 8y + 12$$

Algebraic expressions are tested to be equivalent by choosing values for the variables and evaluating both expressions (see 2.A.4). For example, $4(3x - 2)$ and $12x - 8$ are tested by substituting 3 for the variable *x* and calculating to determine if equivalent values result.

Evaluating Expressions for Given Values

An algebraic expression is a statement written in mathematical symbols, typically including one or more unknown values represented by variables. For example, the expression $2x + 3$ states that an unknown number (*x*) is multiplied by 2 and added to 3. If given a value for the unknown number, or variable, the value of the expression is determined. For example, if the value of the variable *x* is 4, the value of the expression 4 is multiplied by 2, and 3 is added. This results in a value of 11 for the expression.

When given an algebraic expression and values for the variable(s), the expression is evaluated to determine its numerical value. To evaluate the expression, the given values for the variables are substituted (or replaced) and the expression is simplified using the order of operations. Parenthesis should be used when substituting. Consider the following: Evaluate $a - 2b + ab$ for $a = 3$ and $b = -1$. To evaluate, any variable *a* is replaced with 3 and any variable *b* with -1, producing:

$$3 - 2(-1) + 3(-1)$$

Next, the order of operations is used to calculate the value of the expression, which is 2.

Verbal Statements and Algebraic Expressions

As mentioned, an algebraic expression is a statement about unknown quantities expressed in mathematical symbols. The statement *five times a number added to forty* is expressed as $5x + 40$. An equation is a statement in which two expressions (with at least one containing a variable) are equal to one another. The statement *five times a number added to forty is equal to ten* is expressed as:

$$5x + 40 = 10$$

Real world scenarios can also be expressed mathematically. Suppose a job pays its employees $300 per week and $40 for each sale made. The weekly pay is represented by the expression $40x + 300$ where *x* is the number of sales made during the week.

Consider the following scenario: Bob had $20 and Tom had $4. After selling 4 ice cream cones to Bob, Tom has as much money as Bob. The cost of an ice cream cone is an unknown quantity and can be represented by a variable (*x*). The amount of money Bob has after his purchase is four times the cost of an ice cream cone subtracted from his original:

$$\$20 \to 20 - 4x$$

The amount of money Tom has after his sale is four times the cost of an ice cream cone added to his original:

$$\$4 \to 4x + 4$$

After the sale, the amount of money that Bob and Tom have are equal:

$$\to 20 - 4x = 4x + 4$$

Solving for x yields $x = 2$.

Use of Formulas

Formulas are mathematical expressions that define the value of one quantity, given the value of one or more different quantities. Formulas look like equations because they contain variables, numbers, operators, and an equal sign. All formulas are equations but not all equations are formulas. A formula must have more than one variable. For example:

$$2x + 7 = y$$

is an equation and a formula (it relates the unknown quantities *x* and *y*). However:

$$2x + 7 = 3$$

is an equation but not a formula (it only expresses the value of the unknown quantity *x*).

Formulas are typically written with one variable alone (or isolated) on one side of the equal sign. This variable can be thought of as the *subject* in that the formula is stating the value of the *subject* in terms of the relationship between the other variables. Consider the distance formula: $distance = rate \times time$ or $d = rt$. The value of the subject variable *d* (distance) is the product of the variable *r* and *t* (rate and time). Given the rate and time, the distance traveled can easily be determined by substituting the values into the formula and evaluating.

The formula:

$$P = 2l + 2w$$

expresses how to calculate the perimeter of a rectangle (*P*) given its length (*l*) and width (*w*). To find the perimeter of a rectangle with a length of 3ft and a width of 2ft, these values are substituted into the formula for *l* and *w*:

$$P = 2(3ft) + 2(2ft)$$

Following the order of operations, the perimeter is determined to be 10ft. When working with formulas such as these, including units is an important step.

Given a formula expressed in terms of one variable, the formula can be manipulated to express the relationship in terms of any other variable. In other words, the formula can be rearranged to change which variable is the **subject.** To solve for a variable of interest by manipulating a formula, the equation may be solved as if all other variables were numbers. The same steps for solving are followed, leaving operations in terms of the variables instead of calculating numerical values. For the formula:

$$P = 2l + 2w$$

the perimeter is the subject expressed in terms of the length and width. To write a formula to calculate the width of a rectangle, given its length and perimeter, the previous formula relating the three variables is solved for the variable *w*. If *P* and *l* were numerical values, this is a two-step linear equation solved by subtraction and division. To solve the equation $P = 2l + 2w$ for *w*, 2*l* is first subtracted from both sides:

$$P - 2l = 2w$$

Then both sides are divided by 2:

$$\frac{P - 2l}{2} = w$$

Dependent and Independent Variables

A **variable** represents an unknown quantity and, in the case of a formula, a specific relationship exists between the variables. Within a given scenario, variables are the quantities that are changing. If two variables exist, one is dependent and one is independent. The value of one variable depends on the other variable. If a scenario describes distance traveled and time traveled at a given speed, distance is dependent and time is independent. The distance traveled depends on the time spent traveling. If a scenario describes the cost of a cab ride and the distance traveled, the cost is dependent and the distance is independent. The cost of a cab ride depends on the distance travelled. Formulas often contain more than two variables and are typically written with the dependent variable alone on one side of the equation. This lone variable is the *subject* of the statement. If a formula contains three or more variables, one variable is dependent and the rest are independent. The values of all independent variables are needed to determine the value of the dependent variable.

The formula $C = 2\pi r$ expresses the dependent variable *C,* the circumference of a circle, in terms of the independent variables, *r*—the radius. The circumference of a circle depends on its radius. The formula

$$d = rt\ (distance = rate \times time)$$

expresses the dependent variable *d* in terms of the independent variables, *r* and *t*. The distance traveled depends on the rate (or speed) and the time traveled.

Solving Simple Algebraic Problems

Linear equations and **linear inequalities** are both comparisons of two algebraic expressions. However, unlike equations in which the expressions are equal, linear inequalities compare expressions that may be unequal. Linear equations typically have one value for the variable that makes the statement true. Linear inequalities generally have an infinite number of values that make the statement true.

When solving a linear equation, the desired result requires determining a numerical value for the unknown **variable**. If given a linear equation involving addition, subtraction, multiplication, or division, working backwards isolates the variable. Addition and subtraction are inverse operations, as are multiplication and division. Therefore, they can be used to cancel each other out.

Since variables are the letters that represent an unknown number, you must solve for that unknown number in single variable problems. The main thing to remember is that you can do anything to one side of an equation as long as you do it to the other.

The first steps to solving linear equations are distributing, if necessary, and combining any like terms on the same side of the equation. Sides of an equation are separated by an **equal sign**. Next, the equation is manipulated to show the variable on one side. Again, whatever is done to one side of the equation must be done to the other side of the equation to remain equal. Inverse operations are then used to isolate the variable and undo the order of operations backwards. Addition and subtraction are undone, then multiplication and division are undone.

For example, solve $4(t-2)+2t-4=2(9-2t)$

Distribute: $4t-8+2t-4=18-4t$

Combine like terms: $6t-12=18-4t$

Add $4t$ to each side to move the variable: $10t-12=18$

Add 12 to each side to isolate the variable: $10t=30$

Divide each side by 10 to isolate the variable: $t=3$

The answer can be checked by substituting the value for the variable into the original equation, ensuring that both sides calculate to be equal.

Linear inequalities express the relationship between unequal values. More specifically, they describe in what way the values are unequal. A value can be greater than (>), less than (<), greater than or equal to (≥), or less than or equal to (≤) another value.

$$5x+40>65$$

is read as *five times a number added to forty is greater than sixty-five.*

When solving a linear inequality, the solution is the set of all numbers that make the statement true. The inequality $x+2\geq 6$ has a solution set of 4 and every number greater than 4 (4.01; 5; 12; 107; etc.). Adding 2 to 4 or any number greater than 4 results in a value that is greater than or equal to 6. Therefore, $x\geq 4$ is the solution set.

To algebraically solve a linear inequality, follow the same steps as those for solving a linear equation. The inequality symbol stays the same for all operations except when multiplying or dividing by a negative number. If multiplying or dividing by a negative number while solving an inequality, the relationship reverses (the sign flips). In other words, > switches to < and vice versa. Multiplying or dividing by a positive number does not change the relationship, so the sign stays the same.

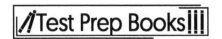

An example is shown below.

Solve $-2x - 8 \leq 22$

Add 8 to both sides: $-2x \leq 30$

Divide both sides by -2: $x \geq -15$

Although linear equations generally have one solution, this is not always the case. If there is no value for the variable that makes the statement true, there is no solution to the equation. Consider the equation:

$$x + 3 = x - 1$$

There is no value for *x* in which adding 3 to the value produces the same result as subtracting one from the value. Conversely, if any value for the variable makes a true statement, the equation has an infinite number of solutions. Consider the equation:

$$3x + 6 = 3(x + 2)$$

Any number substituted for *x* will result in a true statement (both sides of the equation are equal).

By manipulating equations like the two above, the variable of the equation will cancel out completely. If the remaining constants express a true statement (ex. $6 = 6$), then all real numbers are solutions to the equation. If the constants left express a false statement (ex. $3 = -1$), then no solution exists for the equation.

When solving radical and rational equations, extraneous solutions must be accounted for when finding the answers. For example, the equation:

$$\frac{x}{x-5} = \frac{3x}{x+3}$$

has two values that create a 0 denominator: $x \neq 5, -3$. When solving for x, these values must be considered because they cannot be solutions. In the given equation, solving for x can be done using cross-multiplication, yielding the equation:

$$x(x + 3) = 3x(x - 5)$$

Distributing results in the quadratic equation yields:

$$x^2 + 3x = 3x^2 - 15x$$

Therefore, all terms must be moved to one side of the equals sign. This results in:

$$2x^2 - 18x = 0$$

which in factored form is:

$$2x(x - 9) = 0$$

Setting each factor equal to zero, the apparent solutions are $x = 0$ and $x = 9$. These two solutions are neither 5 nor -3, so they are viable solutions. Neither 0 nor 9 create a 0 denominator in the original equation.

A similar process exists when solving radical equations. One must check to make sure the solutions are defined in the original equations. Solving an equation containing a square root involves isolating the root and then squaring both sides of the equals sign. Solving a cube root equation involves isolating the radical and then cubing both sides. In either case, the variable can then be solved for because there are no longer radicals in the equation.

Solving a linear inequality requires all values that make the statement true to be determined. For example, solving:

$$3x - 7 \geq -13$$

produces the solution $x \geq -2$. This means that -2 and any number greater than -2 produces a true statement. Solution sets for linear inequalities will often be displayed using a number line. If a value is included in the set (≥ or ≤), a shaded dot is placed on that value and an arrow extending in the direction of the solutions. For a variable > or ≥ a number, the arrow will point right on a number line, the direction where the numbers increase. If a variable is < or ≤ a number, the arrow will point left on a number line, which is the direction where the numbers decrease. If the value is not included in the set (> or <), an open (unshaded) circle on that value is used with an arrow in the appropriate direction.

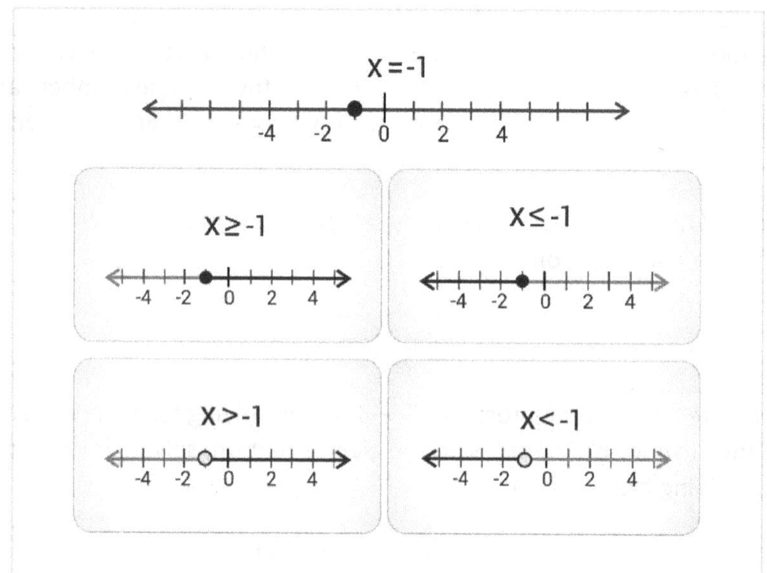

Similar to linear equations, a linear inequality may have a solution set consisting of all real numbers, or can contain no solution. When solved algebraically, a linear inequality in which the variable cancels out and results in a true statement (ex. $7 \geq 2$) has a solution set of all real numbers. A linear inequality in which the variable cancels out and results in a false statement (ex. $7 \leq 2$) has no solution.

Equations and inequalities in two variables represent a relationship. Jim owns a car wash and charges $40 per car. The rent for the facility is $350 per month. An equation can be written to relate the number of cars Jim cleans to the money he makes per month. Let x represent the number of cars and y represent the profit Jim makes each month from the car wash. The equation:

$$y = 40x - 350$$

can be used to show Jim's profit or loss. Since this equation has two variables, the coordinate plane can be used to show the relationship and predict profit or loss for Jim. The following graph shows that Jim must wash at least nine cars to pay the rent, where $x = 9$. Anything nine cars and above yield a profit shown in the value on the y-axis.

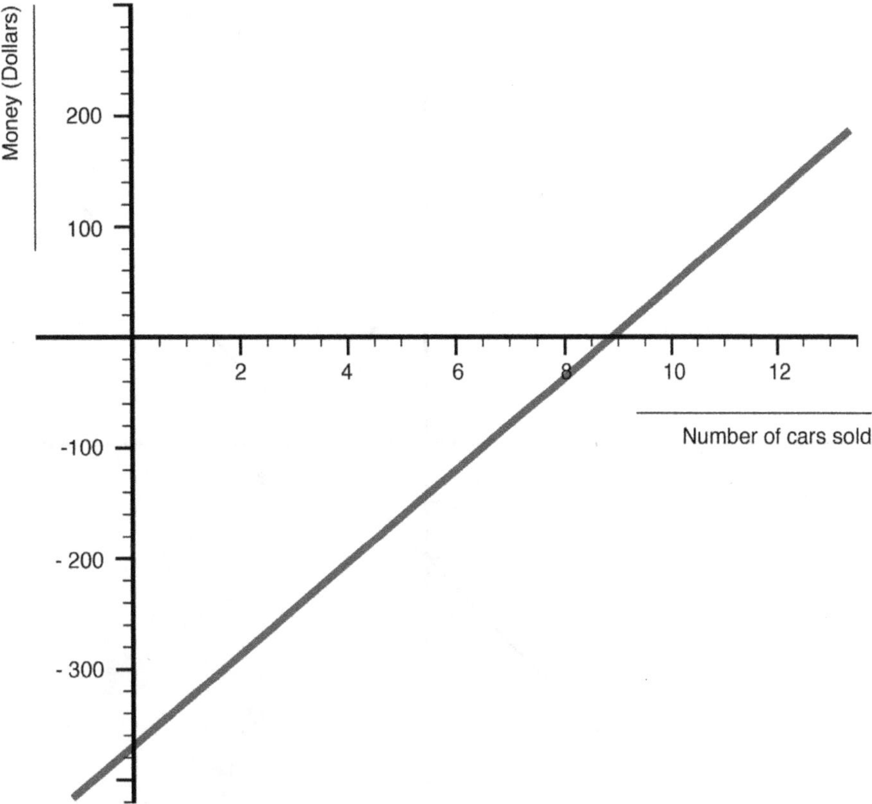

With a single equation in two variables, the solutions are limited only by the situation the equation represents. When two equations or inequalities are used, more constraints are added. For example, in a system of linear equations, there is often—although not always—only one answer. The point of intersection of two lines is the solution. For a system of inequalities, there are infinitely many answers.

The intersection of two solution sets gives the solution set of the system of inequalities. In the following graph, the darker shaded region is where two inequalities overlap. Any set of x and y found in that region satisfies both inequalities. The line with the positive slope is solid, meaning the values on that line are included in the solution. The line with the negative slope is dotted, so the coordinates on that line are not included.

Here's an example:

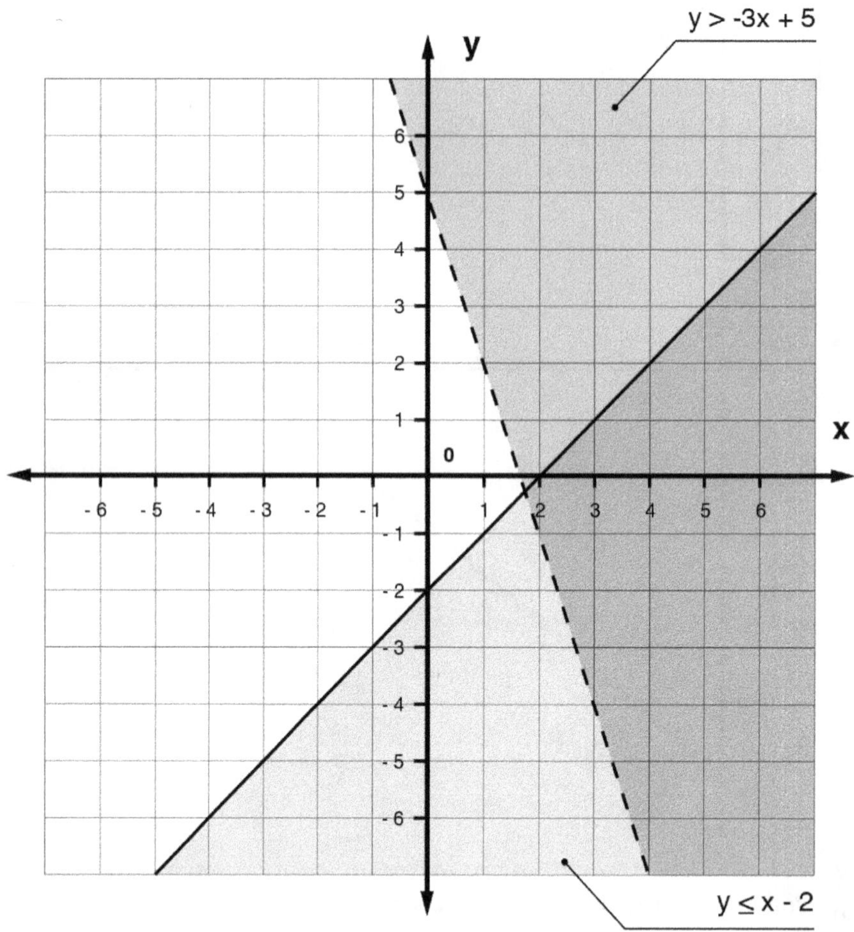

Formulas with two variables are equations used to represent a specific relationship. For example, the formula $d = rt$ represents the relationship between distance, rate, and time. If Bob travels at a rate of 35 miles per hour on his road trip from Westminster to Seneca, the formula $d = 35t$ can be used to represent his distance traveled in a specific length of time. Formulas can also be used to show different roles of the variables, transformed without any given numbers. Solving for r, the formula becomes $\frac{d}{t} = r$. The t is moved over by division so that rate is a function of distance and time.

Solving Equations

Solving equations in one variable involves isolating that variable on one side of the equation. The letters in an equation and any numbers attached to them are the variables as they stand for unknown quantities that you are trying to solve for. For example, in $3x - 7 = 20$, the variable x needs to be isolated. Using opposite operations, the -7 is moved to the right side of the equation by adding seven to both sides:

$$3x - 7 + 7 = 20 + 7$$

$$3x = 27$$

Dividing by three on each side:

$$\frac{3x}{3} = \frac{27}{3}$$

isolates the variable. It is important to note that if an operation is performed on one side of the equals sign, it has to be performed on the other side to maintain equality. The solution is found to be $x = 9$. This solution can be checked for accuracy by plugging x=9 in the original equation. After simplifying the equation, $20 = 20$ is found, which is a true statement.

When solving radical and rational equations, extraneous solutions must be accounted for when finding the answers. For example, the equation:

$$\frac{x}{x-5} = \frac{3x}{x+3}$$

has two values that create a 0 denominator: $x \neq 5, -3$. When solving for x, these values must be considered because they cannot be solutions. In the given equation, solving for x can be done using cross-multiplication, yielding the equation:

$$x(x+3) = 3x(x-5)$$

Distributing results in the quadratic equation yields:

$$x^2 + 3x = 3x^2 - 15x$$

therefore, all terms must be moved to one side of the equals sign. This results in:

$$2x^2 - 18x = 0$$

which in factored form is:

$$2x(x-9) = 0$$

Setting each factor equal to zero, the apparent solutions are $x = 0$ and $x = 9$. These two solutions are neither 5 nor -3, so they are viable solutions. Neither 0 nor 9 create a 0 denominator in the original equation.

A similar process exists when solving radical equations. One must check to make sure the solutions are defined in the original equations. Solving an equation containing a square root involves isolating the root and then squaring both sides of the equals sign. Solving a cube root equation involves isolating the radical and then cubing both sides. In either case, the variable can then be solved for because there are no longer radicals in the equation.

Equations with one variable can be solved using the addition principle and multiplication principle. If $a = b$, then $a + c = b + c$, and $ac = bc$. Given the equation:

$$2x - 3 = 5x + 7$$

the first step is to combine the variable terms and the constant terms. Using the principles, expressions can be added and subtracted onto and off both sides of the equals sign, so the equation turns into

$-10 = 3x$. Dividing by 3 on both sides through the multiplication principle with $c = \frac{1}{3}$ results in the final answer of $x = \frac{-10}{3}$.

Some equations have a higher degree and are not solved by simply using opposite operations. When an equation has a degree of 2, completing the square is an option. For example, the quadratic equation:

$$x^2 - 6x + 2 = 0$$

can be rewritten by completing the square. A **quadratic equation** is an equation in the form:

$$ax^2 + bx + c = 0$$

The goal of completing the square is to get the equation into the form:

$$(x - p)^2 = q$$

Using the example, the constant term 2 first needs to be moved over to the opposite side by subtracting. Then, the square can be completed by adding 9 to both sides, which is the square of half of the coefficient of the middle term $-6x$. The current equation is:

$$x^2 - 6x + 9 = 7$$

The left side can be factored into a square of a binomial, resulting in:

$$(x - 3)^2 = 7$$

To solve for x, the square root of both sides should be taken, resulting in:

$$(x - 3) = \pm\sqrt{7}$$

$$x = 3 \pm \sqrt{7}$$

Other ways of solving quadratic equations include graphing, factoring, and using the quadratic formula. The equation:

$$y = x^2 - 4x + 3$$

can be graphed on the coordinate plane, and the solutions can be observed where the graph crosses the x-axis. The graph will be a parabola that opens up with two solutions at 1 and 3.

The equation can also be factored to find the solutions. The original equation:

$$y = x^2 - 4x + 3$$

can be factored into:

$$y = (x - 1)(x - 3)$$

Setting this equal to zero, the x-values are found to be 1 and 3, just as on the graph. Solving by factoring and graphing are not always possible. The **quadratic formula** is a method of solving quadratic equations that always results in exact solutions.

The formula is:

$$x = \frac{-b \pm \sqrt{b^2 - 4ac}}{2a}$$

where a, b, and c are the coefficients in the original equation in standard form:

$$y = ax^2 + bx + c$$

For this example,

$$x = \frac{4 \pm \sqrt{(-4)^2 - 4(1)(3)}}{2(1)}$$

$$\frac{4 \pm \sqrt{16 - 12}}{2}$$

$$\frac{4 \pm 2}{2}$$

$$1, 3$$

Multistep One-Variable Linear Equations and Inequalities

Solutions of a linear equation or a linear inequality are the values of the variable that make a statement true. In the case of a linear equation, the solution set (list of all possible solutions) typically consists of a single numerical value. To find the solution, the equation is solved by isolating the variable. For example, solving the equation $3x - 7 = -13$ produces the solution $x = -2$. The only value for *x* which produces a true statement is -2. This can be checked by substituting -2 into the original equation to check that both sides are equal. In this case:

$$3(-2) - 7 = -13 \rightarrow -13 = -13$$

Therefore, -2 is a solution.

Although linear equations generally have one solution, this is not always the case. If there is no value for the variable that makes the statement true, there is no solution to the equation. Consider the equation:

$$x + 3 = x - 1$$

There is no value for *x* in which adding 3 to the value produces the same result as subtracting one from the value. Conversely, if any value for the variable makes a true statement, the equation has an infinite number of solutions. Consider the equation:

$$3x + 6 = 3(x + 2)$$

Any number substituted for *x* will result in a true statement (both sides of the equation are equal).

By manipulating equations like the two above, the variable of the equation will cancel out completely. If the remaining constants express a true statement (ex. $6 = 6$), then all real numbers are solutions to the equation. If the constants left express a false statement (ex. $3 = -1$), then no solution exists for the equation.

Linear Relationships

Linear relationships describe the way two quantities change with respect to each other. The relationship is defined as **linear** because a line is produced if all the sets of corresponding values are graphed on a coordinate grid. When expressing the linear relationship as an equation, the equation is often written in the form:

$$y = mx + b \text{ (slope-intercept form)}$$

where *m* and *b* are numerical values and *x* and *y* are variables (for example, $y = 5x + 10$). Given a linear equation and the value of either variable (*x* or *y*), the value of the other variable can be determined.

Suppose a teacher is grading a test containing 20 questions with 5 points given for each correct answer, adding a curve of 10 points to each test. This linear relationship can be expressed as the equation:

$$y = 5x + 10$$

where *x* represents the number of correct answers and *y* represents the test score. To determine the score of a test with a given number of correct answers, the number of correct answers is substituted into the equation for *x* and evaluated. For example, for 10 correct answers, 10 is substituted for *x*:

$$y = 5(10) + 10 \rightarrow y = 60$$

Therefore, 10 correct answers will result in a score of 60. The number of correct answers needed to obtain a certain score can also be determined. To determine the number of correct answers needed to score a 90, 90 is substituted for *y* in the equation (*y* represents the test score) and solved:

$$90 = 5x + 10 \rightarrow 80 = 5x \rightarrow 16 = x$$

Therefore, 16 correct answers are needed to score a 90.

Linear relationships may be represented by a table of 2 corresponding values. Certain tables may determine the relationship between the values and predict other corresponding sets. Consider the table below, which displays the money in a checking account that charges a monthly fee:

Month	0	1	2	3	4
Balance	$210	$195	$180	$165	$150

An examination of the values reveals that the account loses $15 every month (the month increases by one and the balance decreases by 15). This information can be used to predict future values. To determine what the value will be in month 6, the pattern can be continued, and it can be concluded that the balance will be $120. To determine which month the balance will be $0, $210 is divided by $15 (since the balance decreases $15 every month), resulting in month 14.

Similar to a table, a graph can display corresponding values of a linear relationship.

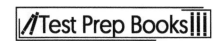

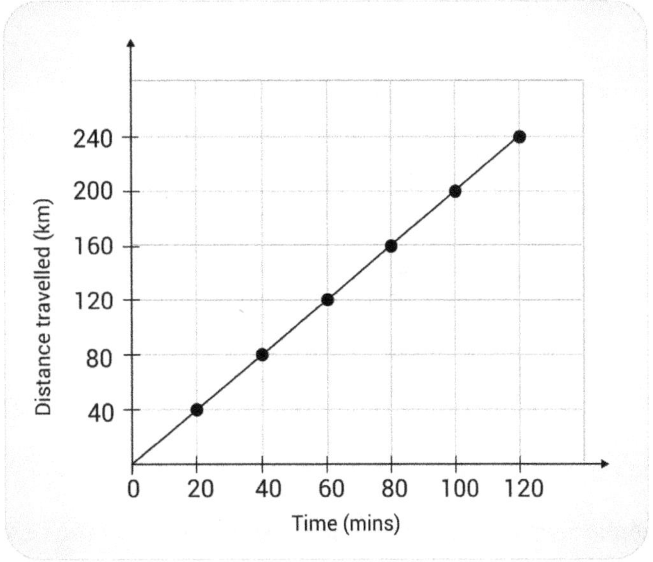

The graph above represents the relationship between distance traveled and time. To find the distance traveled in 80 minutes, the mark for 80 minutes is located at the bottom of the graph. By following this mark directly up on the graph, the corresponding point for 80 minutes is directly across from the 160 kilometer mark. This information indicates that the distance travelled in 80 minutes is 160 kilometers. To predict information not displayed on the graph, the way in which the variables change with respect to one another is determined. In this case, distance increases by 40 kilometers as time increases by 20 minutes. This information can be used to continue the data in the graph or convert the values to a table.

Equations and Graphing

As mentioned, a function is called **linear** if it can take the form of the equation:

$$f(x) = ax + b, \text{or } y = ax + b$$

for any two numbers *a* and *b*. A linear equation forms a straight line when graphed on the coordinate plane. An example of a linear function is shown below on the graph.

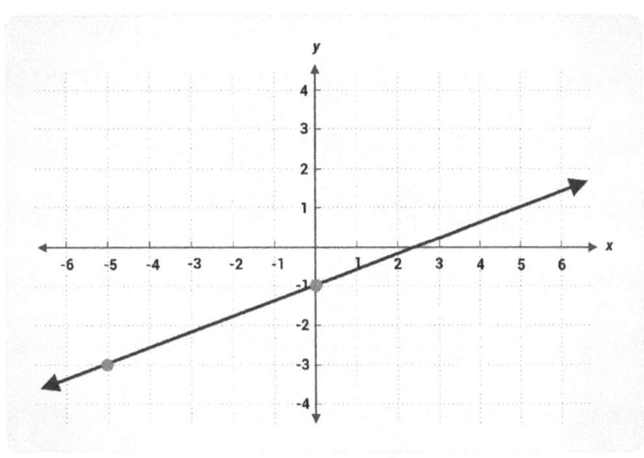

This is a graph of the following function:

$$y = \frac{2}{5}x - 1$$

A table of values that satisfies this function is shown below.

x	y
-5	-3
0	-1
5	1
10	3

These points can be found on the graph using the form (x, y).

To graph relations and functions, the **Cartesian plane** is used. The plane can be visualized as a grid of squares, with one direction being the *x*-axis and the other direction the *y*-axis. Generally, the independent variable is placed along the horizontal (x) axis, and the dependent variable is placed along the vertical (y) axis. Any point on the plane can be specified with a pair of numbers (x, y) that represent how far to go along the *x*-axis and how far to go up or down the *y*-axis. Specific values for these pairs can be given names such as $C = (-1, 3)$. Negative values mean to move left or down; positive values mean to move right or up. The point where the axes cross one another is called the **origin**. The origin has coordinates $(0, 0)$ and is usually called O when given a specific label.

An illustration of the Cartesian plane, along with graphs of $(2, 1)$ and $(-1, -1)$, are below.

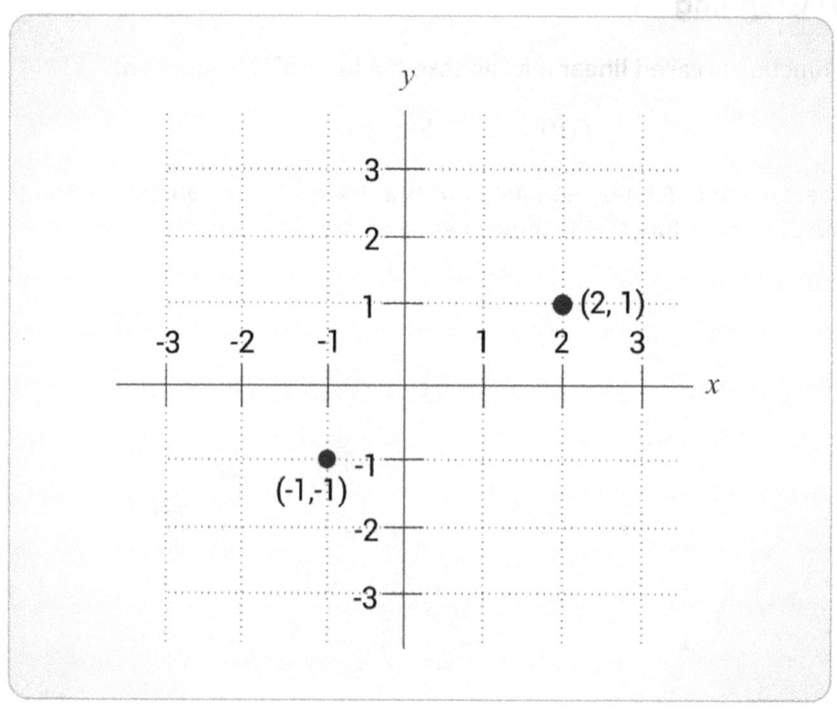

Relations also can be graphed by marking each point whose coordinates satisfy the relation. If the relation is a function, then there is only one value of *y* for any given value of x. This leads to the **vertical line test**: if a relation is graphed, then the relation is a function if any possible vertical line drawn anywhere along the graph would only touch the graph of the relation in no more than one place. Conversely, when graphing a function, then any possible vertical line drawn will not touch the graph of the function at any point or will touch the function at just one point. This test is made from the definition of a function, where each x-value must be mapped to one and only one y-value.

When graphing a linear function, note that the ratio of the change of the *y* coordinate to the change in the *x* coordinate is constant between any two points on the resulting line, no matter which two points are chosen. In other words, in a pair of points on a line, (x_1, y_1) and (x_2, y_2), with $x_1 \neq x_2$ so that the two points are distinct, then the ratio:

$$\frac{y_2 - y_1}{x_2 - x_1}$$

will be the same, regardless of which particular pair of points are chosen. This ratio:

$$\frac{y_2 - y_1}{x_2 - x_1}$$

is called the **slope** of the line and is frequently denoted with the letter m. If slope *m* is positive, then the line goes upward when moving to the right, while if slope *m* is negative, then the line goes downward when moving to the right. If the slope is 0, then the line is called **horizontal,** and the *y*-coordinate is constant along the entire line. In lines where the *x*-coordinate is constant along the entire line, *y* is not actually a function of *x*. For such lines, the slope is not defined. These lines are called **vertical** lines.

Linear functions may take forms other than $y = ax + b$. The most common forms of linear equations are explained below:

1. Standard Form: $Ax + By = C$, in which the slope is given by $m = \frac{-A}{B}$, and the *y*-intercept is given by $\frac{C}{B}$.

2. Slope-Intercept Form: $y = mx + b$, where the slope is *m* and the *y* intercept is *b*.

3. Point-Slope Form: $y - y_1 = m(x - x_1)$, where the slope is *m* and (x_1, y_1) is any point on the chosen line.

4. Two-Point Form:

$$\frac{y - y_1}{x - x_1} = \frac{y_2 - y_1}{x_2 - x_1}$$

where (x_1, y_1) and (x_2, y_2) are any two distinct points on the chosen line. Note that the slope is given by:

$$m = \frac{y_2 - y_1}{x_2 - x_1}$$

5. Intercept Form: $\frac{x}{x_1} + \frac{y}{y_1} = 1$, in which x_1 is the *x*-intercept and y_1 is the *y*-intercept.

These five ways to write linear equations are all useful in different circumstances. Depending on the given information, it may be easier to write one of the forms over another.

If $y = mx$, y is directly proportional to x. In this case, changing x by a factor changes y by that same factor. If $y = \frac{m}{x}$, y is inversely proportional to x. For example, if x is increased by a factor of 3, then y will be decreased by the same factor, 3.

The **midpoint** between two points, (x_1, y_1) and (x_2, y_2), is given by taking the average of the x-coordinates and the average of the y-coordinates:

$$\left(\frac{x_1 + x_2}{2}, \frac{y_1 + y_2}{2}\right)$$

The **distance** between two points, (x_1, y_1) and (x_2, y_2), is given by the **Pythagorean formula**:

$$\sqrt{(x_2 - x_1)^2 + (y_2 - y_1)^2}$$

To find the perpendicular distance between a line $Ax + By = C$ and a point (x_1, y_1) not on the line, we need to use the formula:

$$\frac{|Ax_1 + By_1 + C|}{\sqrt{A^2 + B^2}}$$

Functions

A **function** is defined as a relationship between inputs and outputs where there is only one output value for a given input. The input is called the **independent variable**. If the variable is set equal to the output, as in $y = f(x)$, then this is called the **dependent variable**. To indicate the dependent value a function, y, gives for a specific independent variable, x, the notation y = $f(x)$ is used.

As an example, the following function is in function notation:

$$f(x) = 3x - 4$$

The $f(x)$ represents the output value for an input of x. If $x = 2$, the equation becomes:

$$f(2) = 3(2) - 4 = 6 - 4 = 2$$

The **input** of 2 yields an **output** of 2, forming the ordered pair $(2, 2)$. The following set of ordered pairs corresponds to the given function:

$$(2, 2), (0, -4), (-2, -10)$$

The set of all possible inputs of a function is its **domain**, and all possible outputs is called the **range.** By definition, each member of the domain is paired with only one member of the range.

Functions can also be defined recursively. In this form, they are not defined explicitly in terms of variables. Instead, they are defined using previously-evaluated function outputs, starting with either $f(0)$ or $f(1)$. An example of a recursively-defined function is:

$$f(1) = 2, f(n) = 2f(n - 1) + 2n, n > 1$$

The domain of this function is the set of all integers.

The domain and range of a function can be found visually by its plot on the coordinate plane. In the function:

$$f(x) = x^2 - 3$$

for example, the domain is all real numbers because the parabola stretches as far left and as far right as it can go, with no restrictions. This means that any input value from the real number system will yield an answer in the real number system. For the range, the inequality $y \geq -3$ would be used to describe the possible output values because the parabola has a minimum at $y = -3$. This means there will not be any real output values less than -3 because -3 is the lowest value it reaches on the y-axis.

These same answers for domain and range can be found by observing a table. The table below shows that from input values $x = -1$ to $x = 1$, the output results in a minimum of -3. On each side of $x = 0$, the numbers increase, showing that the range is all real numbers greater than or equal to -3.

x (domain/input)	y (range/output)
-2	1
-1	-2
0	-3
-1	-2
2	1

Different types of functions behave in different ways. A function is defined to be increasing over a subset of its domain if for all:

$$x_1 \geq x_2$$

in that interval:

$$f(x_1) \geq f(x_2)$$

Also, a function is decreasing over an interval if for all:

$$x_1 \geq x_2$$

in that interval:

$$f(x_1) \leq f(x_2)$$

A point in which a function changes from increasing to decreasing can also be labeled as the **maximum value** of a function if it is the largest point the graph reaches on the y-axis. A point in which a function changes from decreasing to increasing can be labeled as the **minimum value** of a function if it is the smallest point the graph reaches on the y-axis. Maximum values are also known as **extreme values**. The graph of a **continuous function** does not have any breaks or jumps in the graph. This description is not true of all functions. A **radical function**, for example:

$$f(x) = \sqrt{x}$$

has a restriction for the domain and range because there are no real negative inputs or outputs for this function. The domain can be stated as $x \geq 0$, and the range is $y \geq 0$.

A **piecewise-defined function** also has a different appearance on the graph. In the following function, there are three equations defined over different intervals. It is a function because there is only one y-value for each x-value, passing the **Vertical Line Test**. The domain is all real numbers less than or equal to 6. The range is all real numbers greater than zero. From left to right, the graph decreases to zero, then increases to almost 4, and then jumps to 6.

From input values greater than 2, the input decreases just below 8 to 4, and then stops.

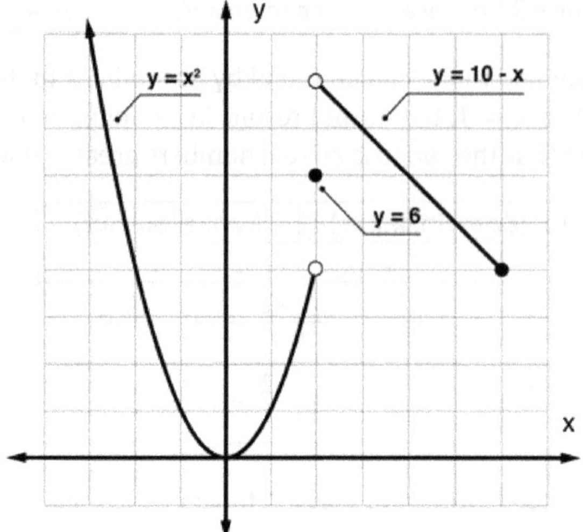

Logarithmic and **exponential functions** also have different behavior than other functions. These two types of functions are inverses of each other. The **inverse** of a function can be found by switching the place of x and y, and solving for y. When this is done for the exponential equation, $y = 2^x$, the function $y = \log_2 x$ is found. The general form of a logarithmic function is $y = log_b x$, which says b raised to the y power equals x.

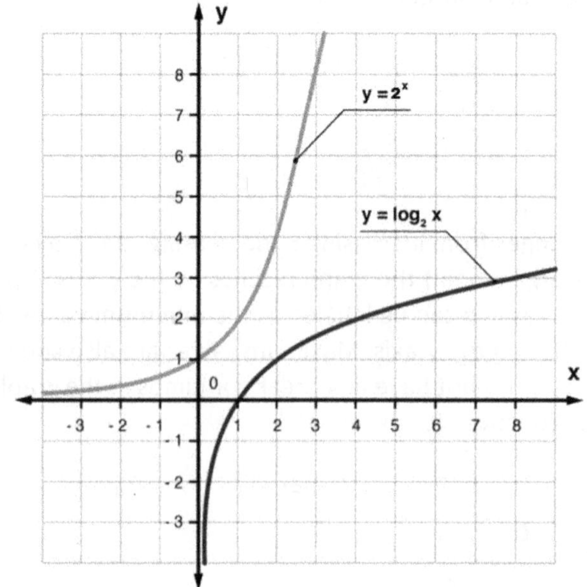

The black line on the graph above represents the logarithmic function $y = \log_2 x$. This curve passes through the point $(1, 0)$, just as all log functions do, because any value $b^0 = 1$. The graph of this logarithmic function starts very close to zero, but does not touch the y-axis. The output value will never be zero by the definition of logarithms. The lighter gray line seen above represents the exponential function $y = 2^x$. The behavior of this function is opposite the logarithmic function because the graph of an inverse function is the graph of the original function flipped over the line $y = x$. The curve passes through the point $(0, 1)$ because any number raised to the zero power is one. This curve also gets very close to the x-axis but never touches it because an exponential expression never has an output of zero. The x-axis on this graph is called a horizontal asymptote. An **asymptote** is a line that represents a boundary for a function. It shows a value that the function will get close to, but never reach.

Functions can also be described as being even, odd, or neither. If $f(-x) = f(x)$ the function is **even**. For example, the function $f(x) = x^2 - 2$ is even. Plugging in $x = 2$ yields an output of $y = 2$. After changing the input to $x = -2$, the output is still $y = 2$. The output is the same for opposite inputs. Another way to observe an even function is by the symmetry of the graph. If the graph is symmetrical about the axis, then the function is even. If the graph is symmetric about the origin, then the function is **odd**.

Algebraically, if $f(-x) = -f(x)$, the function is odd.

Also, a function can be described as **periodic** if it repeats itself in regular intervals. Common periodic functions are trigonometric functions. For example, $y = \sin x$ is a periodic function with period 2π because it repeats itself every 2π units along the x-axis.

Building Functions

Functions can be built out of the context of a situation. For example, the relationship between the money paid for a gym membership and the months that someone has been a member can be described through a function. If the one-time membership fee is $40 and the monthly fee is $30, then the function can be written:

$$f(x) = 30x + 40$$

The x-value represents the number of months the person has been part of the gym, while the output is the total money paid for the membership. The table below shows this relationship. It is a representation of the function because the initial cost is $40 and the cost increases each month by $30.

x (months)	y (money paid to gym)
0	40
1	70
2	100
3	130

Functions can also be built from existing functions. For example, a given function $f(x)$ can be transformed by adding a constant, multiplying by a constant, or changing the input value by a constant. The new function $g(x) = f(x) + k$ represents a vertical shift of the original function. In $f(x) = 3x - 2$ a vertical shift 4 units up would be:

$$g(x) = 3x - 2 + 4$$

$$3x + 2$$

Multiplying the function times a constant k represents a vertical stretch, based on whether the constant is greater than or less than 1. The following function represents a stretch:

$$g(x) = kf(x)$$

$$4(3x - 2)$$

$$12x - 8$$

Changing the input x by a constant forms the function:

$$g(x) = f(x + k)$$

$$3(x + 4) - 2$$

$$3x + 12 - 2$$

$$3x + 10$$

This represents a horizontal shift to the left 4 units. If $(x - 4)$ was plugged into the function, it would represent a vertical shift.

Common Functions

Three common functions used to model different relationships between quantities are linear, quadratic, and exponential functions. Linear functions are the simplest of the three, and the independent variable x has an exponent of 1. Written in the most common form:

$$y = mx + b$$

the coefficient of x tells how fast the function grows at a constant rate, and the b-value tells the starting point. A quadratic function has an exponent of 2 on the independent variable x. Standard form for this type of function is:

$$y = ax^2 + bx + c$$

and the graph is a parabola. These type functions grow at a changing rate. An exponential function has an independent variable in the exponent $y = ab^x$. The graph of these types of functions is described as **growth** or **decay**, based on whether the base, b, is greater than or less than 1. These functions are different from quadratic functions because the base stays constant. A common base is base *e*.

The following three functions model a linear, quadratic, and exponential function respectively:

$$y = 2x,\ y = x^2, \text{ and } y = 2^x$$

Their graphs are shown below. The first graph, modeling the linear function, shows that the growth is constant over each interval. With a horizontal change of 1, the vertical change is 2. It models a constant positive growth. The second graph shows the quadratic function, which is a curve that is symmetric across the y-axis. The growth is not constant, but the change is mirrored over the axis. The last graph models the exponential function, where the horizontal change of 1 yields a vertical change that increases more and more. The exponential graph gets very close to the x-axis, but never touches it, meaning there is an asymptote there. The y-value can never be zero because the base of 2 can never be raised to an input value that yields an output of zero.

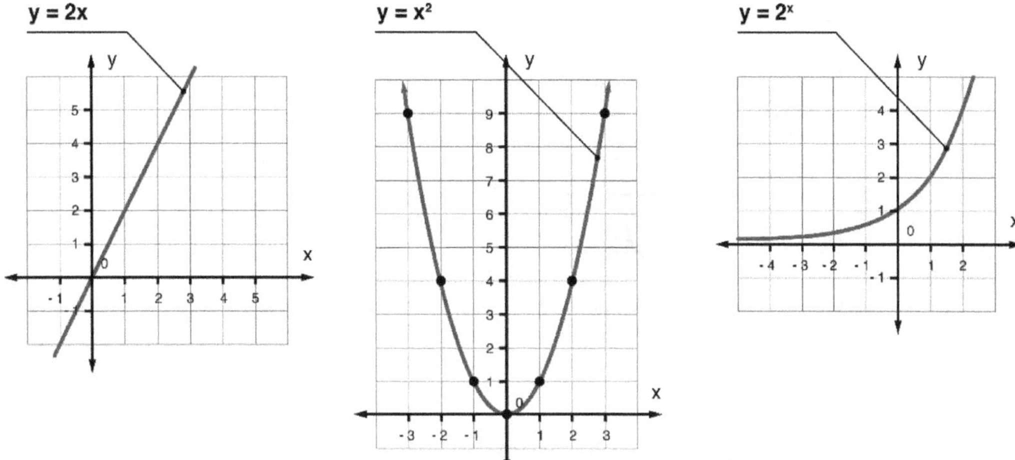

The three tables below show specific values for three types of functions. The third column in each table shows the change in the y-values for each interval. The first table shows a constant change of 2 for each equal interval, which matches the slope in the equation $y = 2x$. The second table shows an increasing change, but it also has a pattern. The increase is changing by 2 more each time, so the change is quadratic. The third table shows the change as factors of the base, 2. It shows a continuing pattern of factors of the base.

y = 2x		
x	y	Δy
1	2	
2	4	2
3	6	2
4	8	2
5	10	2

y = x²		
x	y	Δy
1	1	
2	4	3
3	9	5
4	16	7
5	25	9

y = 2ˣ		
x	y	Δy
1	2	
2	4	2
3	8	4
4	16	8
5	32	16

Given a table of values, the type of function can be determined by observing the change in y over equal intervals. For example, the tables below model two functions. The changes in interval for the x-values is 1 for both tables. For the first table, the y-values increase by 5 for each interval. Since the change is constant, the situation can be described as a linear function. The equation would be:

$$y = 5x + 3$$

For the second table, the change for y is 5, 20, 100, and 500, respectively. The increases are multiples of 5, meaning the situation can be modeled by an exponential function. The equation $y = 5^x + 3$ models this situation.

x	y
0	3
1	8
2	13
3	18
4	23

x	y
0	3
1	8
2	28
3	128
4	628

Quadratic equations can be used to model real-world area problems. For example, a farmer may have a rectangular field that he needs to sow with seed. The field has length $x + 8$ and width $2x$. The formula for area should be used: $A = lw$. Therefore:

$$A = (x + 8) \times 2x = 2x^2 + 16x$$

The possible values for the length and width can be shown in a table, with input x and output A. If the equation was graphed, the possible area values can be seen on the y-axis for given x-values.

Exponential growth and decay can be found in real-world situations. For example, if a piece of notebook paper is folded 25 times, the thickness of the paper can be found. To model this situation, a table can be used. The initial point is one-fold, which yields a thickness of 2 papers. For the second fold, the thickness is 4. Since the thickness doubles each time, the table below shows the thickness for the next few folds. Notice the thickness changes by the same factor each time. Since this change for a constant interval of folds is a factor of 2, the function is exponential. The equation for this is $y = 2^x$. For twenty-five folds, the thickness would be 33,554,432 papers.

x (folds)	y (paper thickness)
0	1
1	2
2	4
3	8
4	16
5	32

Conjectures, Predictions, or Generalizations Based on Patterns

An arithmetic or geometric sequence can be written as a formula and used to determine unknown steps without writing out the entire sequence. An arithmetic sequence progresses by a **common difference**. To determine the common difference, any step is subtracted by the step that precedes it. In the sequence 4, 9, 14, 19 . . . the common difference, or *d*, is 5. By expressing each step as a_1, a_2, a_3, etc., a formula can be written to represent the sequence. a_1 is the first step. To produce step two, step 1 (a_1) is added to the common difference (*d*):

$$a_2 = a_1 + d$$

To produce step three, the common difference (*d*) is added twice to a_1:

$$a_3 = a_1 + 2d$$

To produce step four, the common difference (*d*) is added three times to a_1:

$$a_4 = a_1 + 3d$$

Following this pattern allows a general rule for arithmetic sequences to be written. For any term of the sequence (a_n), the first step (a_1) is added to the product of the common difference (*d*) and one less than the step of the term ($n - 1$):

$$a_n = a_1 + (n - 1)d$$

Suppose the 8th term (a_8) is to be found in the previous sequence. By knowing the first step (a_1) is 4 and the common difference (*d*) is 5, the formula can be used:

$$a_n = a_1 + (n - 1)d$$

$$a_8 = 4 + (7)5$$

$$a_8 = 39$$

In a geometric sequence, each step is produced by multiplying or dividing the previous step by the same number. The **common ratio,** or (*r*), can be determined by dividing any step by the previous step. In the sequence 1, 3, 9, 27 . . . the common ratio (*r*) is:

$$3(\frac{3}{1} = 3 \text{ or } \frac{9}{3} = 3 \text{ or } \frac{27}{9} = 3)$$

Each successive step can be expressed as a product of the first step (a_1) and the common ratio (*r*) to some power. For example:

$$a_2 = a_1 \times r$$

$$a_3 = a_1 \times r \times r \text{ or } a_3 = a_1 \times r^2$$

$$a_4 = a_1 \times r \times r \times r \text{ or } a_4 = a\_1 \times r^3$$

Following this pattern, a general rule for geometric sequences can be written. For any term of the sequence (a_n), the first step (a_1) is multiplied by the common ratio (*r*) raised to the power one less than the step of the term ($n - 1$):

$$a_n = a_1 \times r^{(n-1)}$$

Suppose for the previous sequence, the 7th term (a_7) is to be found. Knowing the first step (a_1) is one, and the common ratio (*r*) is 3, the formula can be used:

$$a_n = a_1 \times r^{(n-1)}$$

$$a_7 = (1) \times 3^6$$

$$a_7 = 729$$

Corresponding Terms of Two Numerical Patterns

When given two numerical patterns, the corresponding terms should be examined to determine if a relationship exists between them. Corresponding terms between patterns are the pairs of numbers that appear in the same step of the two sequences. Consider the following patterns 1, 2, 3, 4 . . . and 3, 6, 9, 12 . . . The corresponding terms are: 1 and 3; 2 and 6; 3 and 9; and 4 and 12. To identify the relationship, each pair of corresponding terms is examined and the possibilities of performing an operation (+, −, ×, ÷) to the term from the first sequence to produce the corresponding term in the second sequence are determined. In this case:

$1 + 2 = 3$ or $1 \times 3 = 3$

$2 + 4 = 6$ or $2 \times 3 = 6$

$3 + 6 = 9$ or $3 \times 3 = 9$

$4 + 8 = 12$ or $4 \times 3 = 12$

The consistent pattern is that the number from the first sequence multiplied by 3 equals its corresponding term in the second sequence. By assigning each sequence a label (input and output) or variable (*x* and *y*), the relationship can be written as an equation. If the first sequence represents the inputs, or *x*, and the second sequence represents the outputs, or *y*, the relationship can be expressed as: $y = 3x$.

Consider the following sets of numbers:

a	2	4	6	8
b	6	8	10	12

To write a rule for the relationship between the values for *a* and the values for *b*, the corresponding terms (2 and 6; 4 and 8; 6 and 10; 8 and 12) are examined. The possibilities for producing *b* from *a* are:

$2 + 4 = 6$ or $2 \times 3 = 6$

$4 + 4 = 8$ or $4 \times 2 = 8$

$6 + 4 = 10$

$8 + 4 = 12$ or $8 \times 1.5 = 12$

The consistent pattern is that adding 4 to the value of *a* produces the value of *b*. The relationship can be written as the equation $a + 4 = b$.

Geometry/Measurement

Geometry deals with shapes and their properties. It is also similar to measurement and number operations. The basis of geometry involves being able to label and describe shapes and their properties. That knowledge will lead to working with formulas such as area, perimeter, and volume. This knowledge will help to solve word problems involving shapes.

Flat or two-dimensional shapes include circles, triangles, hexagons, and rectangles, among others. Three-dimensional solid shapes, such as spheres and cubes, are also used in geometry. A shape can be classified based on whether it is open like the letter U or closed like the letter O. Further classifications involve counting the number of sides and **vertices** (corners) on the shapes. This will help differentiate shapes.

Polygons can be drawn by sketching a fixed number of line segments that meet to create a closed shape. In addition, **triangles** can be drawn by sketching a closed space using only three-line segments. **Quadrilaterals** are closed shapes with four-line segments. Note that a triangle has three vertices, and a quadrilateral has four vertices.

To draw circles, one curved line segment must be drawn that has only one endpoint. This creates a closed shape. Given such direction, every point on the line would be the same distance away from its center. The **radius** of a circle goes from an endpoint on the center of the circle to an endpoint on the circle. The **diameter** is the line segment created by placing an endpoint on the circle, drawing through the radius, and placing the other endpoint on the circle. A compass can be used to draw circles of a more precise size and shape.

Perimeter and Area

Perimeter is the measurement of a distance around something or the sum of all sides of a polygon. Think of perimeter as the length of the boundary, like a fence. In contrast, **area** is the space occupied by a defined enclosure, like the size of a field enclosed by a fence.

When thinking about perimeter, think about walking around the outside of something. When thinking about area, think about the amount of space or **surface area** something takes up.

Squares

The perimeter of a square is measured by adding together all of the sides. Since a square has four equal sides, its perimeter can be calculated by multiplying the length of one side by 4. Thus, the formula is:

$$P = 4 \times s$$

where s equals one side. For example, the following square has side lengths of 5 meters:

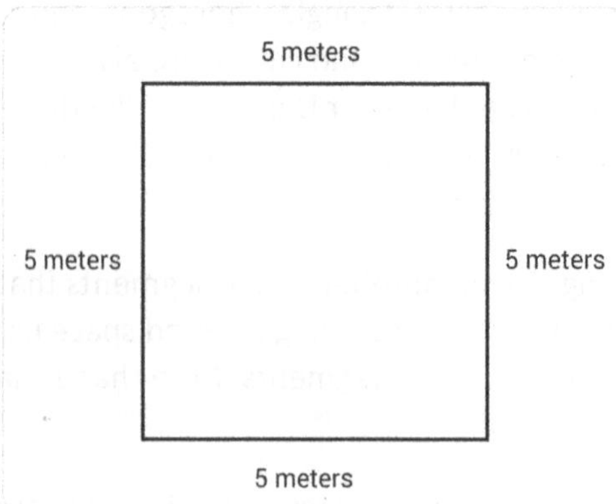

The perimeter is 20 meters because 4 times 5 is 20.

The area of a square is the length of a side squared, and the area of a rectangle is length multiplied by the width. For example, if the length of the square is 7 centimeters, then the area is 49 square centimeters. The formula for this example is:

$$A = s^2 = 7^2 = 49 \text{ square centimeters}$$

Rectangles

Like a square, a rectangle's perimeter is measured by adding together all of the sides. But as the sides are unequal, the formula is different. A rectangle has equal values for its lengths (long sides) and equal values for its widths (short sides), so the perimeter formula for a rectangle is:

$$P = l + l + w + w$$

$$2l + 2w$$

l equals length
w equals width

For example, if the length of a rectangle is 10 inches and the width 8 inches, then the perimeter is 36 inches because:

$$P = 2l + 2w$$

$$2(10) + 2(8)$$

$$20 + 16 = 36 \text{ inches}$$

The area is found by multiplying the length by the width, so the formula is $A = l \times w$.

An example is if the rectangle has a length of 6 inches and a width of 7 inches, then the area is 42 square inches:

$$A = lw = 6(7) = 42 \text{ square inches}$$

Triangles

A triangle's perimeter is measured by adding together the three sides, so the formula is:

$$P = a + b + c$$

where $a, b,$ and c are the values of the three sides. The area is the product of one-half the base and height so the formula is:

$$A = \frac{1}{2} \times b \times h$$

It can be simplified to:

$$A = \frac{bh}{2}$$

The base is the bottom of the triangle, and the height is the distance from the base to the peak. If a problem asks to calculate the area of a triangle, it will provide the base and height.

For example, if the base of the triangle is 2 feet and the height 4 feet, then the area is 4 square feet. The following equation shows the formula used to calculate the area of the triangle:

$$\text{A} = \frac{1}{2}\text{bh} = \frac{1}{2}(2)(4) = 4 \text{ square feet}$$

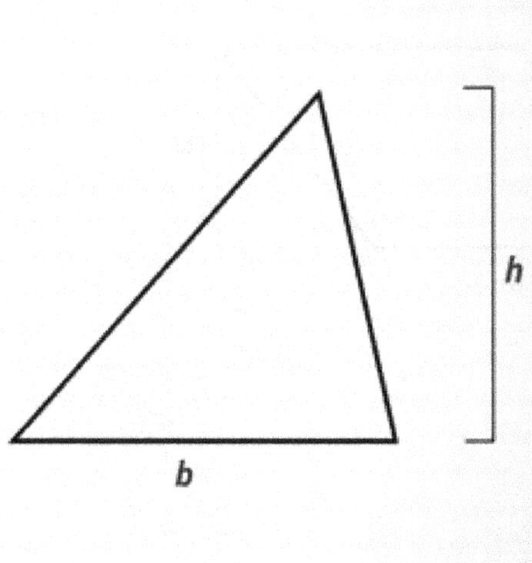

Circles

A circle's perimeter—also known as its **circumference**—is measured by multiplying the diameter by π.

Diameter is the straight line measured from one end to the direct opposite end of the circle.

π is referred to as pi and is equal to 3.14 (with rounding).

So the formula is $\pi \times d$.

This is sometimes expressed by the formula $C = 2 \times \pi \times r$, where r is the radius of the circle. These formulas are equivalent, as the radius equals half of the diameter.

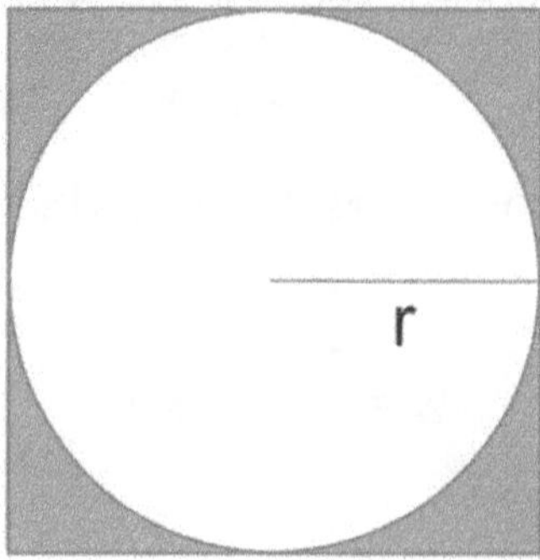

The area of a circle is calculated through the formula

$$A = \pi \times r^2$$

The test will indicate either to leave the answer with π attached or to calculate to the nearest decimal place, which means multiplying by 3.14 for π.

Parallelograms

Similar to triangles, the height of the parallelogram is measured from one base to the other at a 90° angle (or perpendicular).

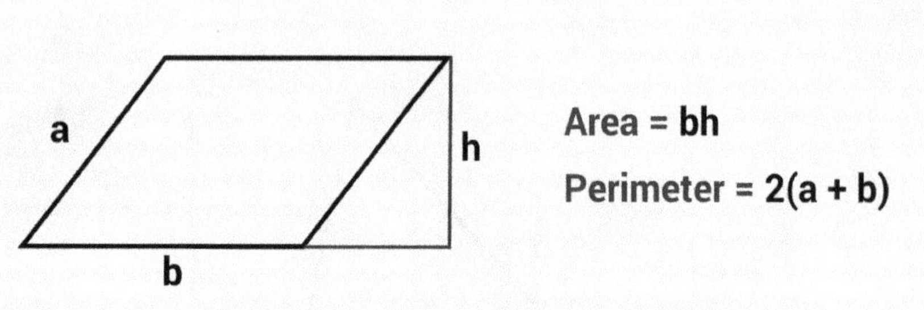

Trapezoid

The area of a trapezoid can be calculated using the formula:

$$A = \frac{1}{2} \times h(b_1 + b_2)$$

where h is the height and b_1 and b_2 are the parallel bases of the trapezoid.

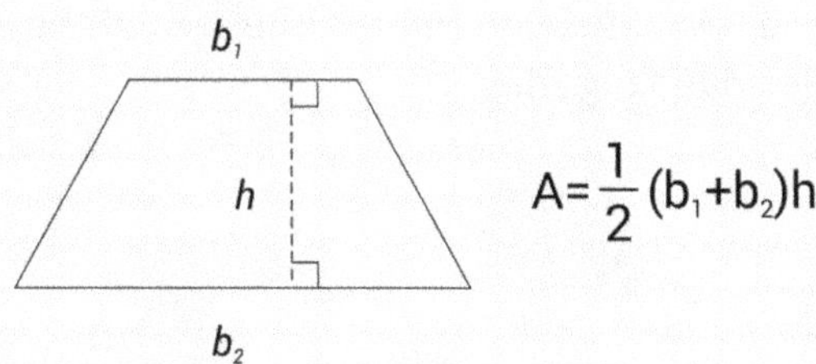

Irregular Shapes

The perimeter of an irregular polygon is found by adding the lengths of all of the sides. In cases where all of the sides are given, this will be very straightforward, as it will simply involve finding the sum of the provided lengths. Other times, a side length may be missing and must be determined before the perimeter can be calculated. Consider the example below:

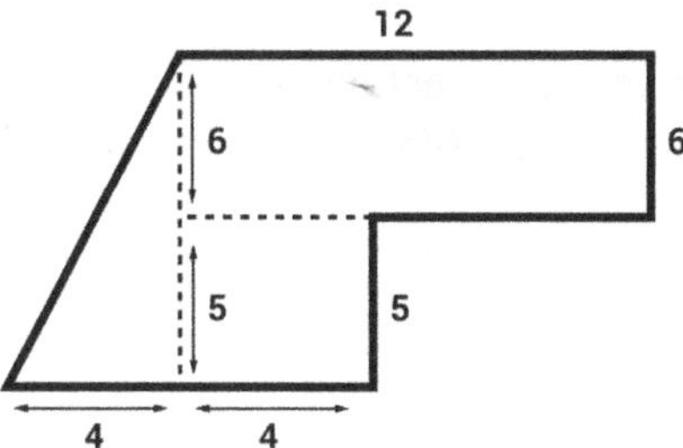

All of the side lengths are provided except for the angled side on the left. Test takers should notice that this is the hypotenuse of a right triangle. The other two sides of the triangle are provided (the base is 4 and the height is 6 + 5 = 11). The Pythagorean Theorem can be used to find the length of the hypotenuse, remembering that $a^2 + b^2 = c^2$.

Substituting the side values provided yields:

$$(4)^2 + (11)^2 = c^2$$

Therefore:

$$c = \sqrt{16 + 121} = 11.7$$

Finally, the perimeter can be found by adding this new side length with the other provided lengths to get the total length around the figure:

$$4 + 4 + 5 + 8 + 6 + 12 + 11.7 = 50.7$$

Although units are not provided in this figure, remember that reporting units with a measurement is important.

The area of irregular polygons is found by decomposing, or breaking apart, the figure into smaller shapes. When the area of the smaller shapes is determined, the area of the smaller shapes will produce the area of the original figure when added together.

Consider the earlier example:

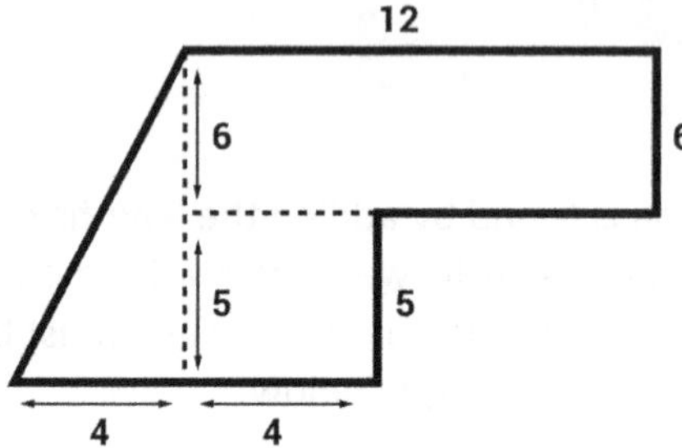

The irregular polygon is decomposed into two rectangles and a triangle. The area of the large rectangles ($A = l \times w \rightarrow A = 12 \times 6$) is 72 square units. The area of the small rectangle is 20 square units:

$$A = 4 \times 5$$

The area of the triangle:

$$A = \frac{1}{2} \times b \times h$$

$$A = \frac{1}{2} \times 4 \times 11 = 22 \; square \; units$$

The sum of the areas of these figures produces the total area of the original polygon:

$$A = 72 + 20 + 22 \rightarrow A = 114 \; square \; units$$

Here's another example:

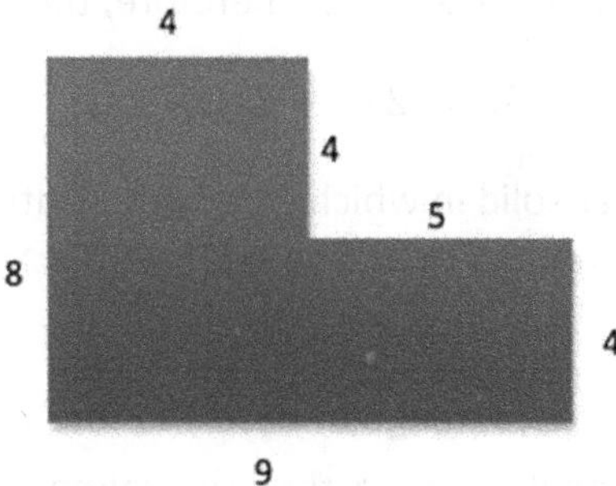

This irregular polygon is decomposed into two rectangles. The area of the large rectangle:

$$A = l \times w$$

$$A = 8 \times 4 = 32 \text{ square units}$$

The area of the small rectangle is 20 square units ($A = 4 \times 5$). The sum of the areas of these figures produces the total area of the original polygon:

$$A = 32 + 20$$

$$A = 52 \text{ square units}$$

Surface Area and Volume

Geometry in three dimensions is similar to geometry in two dimensions. The main new feature is that three points now define a unique **plane** that passes through each of them. Three dimensional objects can be made by putting together two-dimensional figures in different surfaces. Below, some of the possible three-dimensional figures will be provided, along with formulas for their volumes and surface areas.

A **rectangular prism** is a box whose sides are all rectangles meeting at 90° angles. Such a box has three dimensions: length, width, and height. If the length is *x*, the width is *y*, and the height is *z*, then the **volume** is given by $V = xyz$.

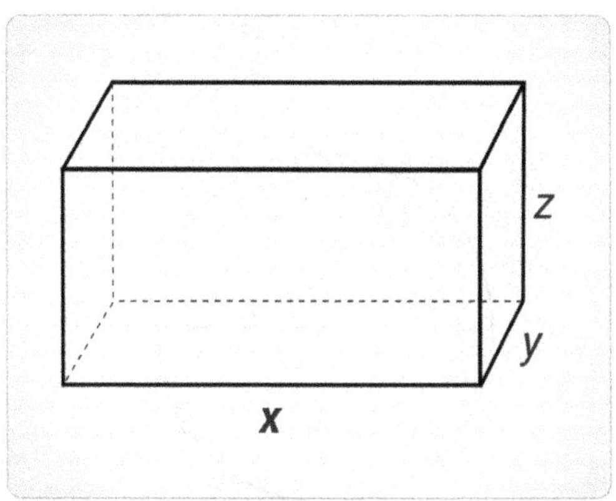

The **surface area** will be given by computing the surface area of each rectangle and adding them together. There are a total of six rectangles. Two of them have sides of length *x* and *y*, two have sides of length *y* and *z*, and two have sides of length *x* and *z*. Therefore, the total surface area will be given by:

$$SA = 2xy + 2yz + 2xz$$

A **cube** is a special type of rectangular solid in which its length, width, and height are the same. If this length is s, then the formula for the volume of a cube is $V = s \times s \times s$. The surface area of a cube is $SA = 6s^2$.

A **rectangular pyramid** is a figure with a rectangular base and four triangular sides that meet at a single vertex. If the rectangle has sides of length *x* and *y*, then the volume will be given by $V = \frac{1}{3}xyh$.

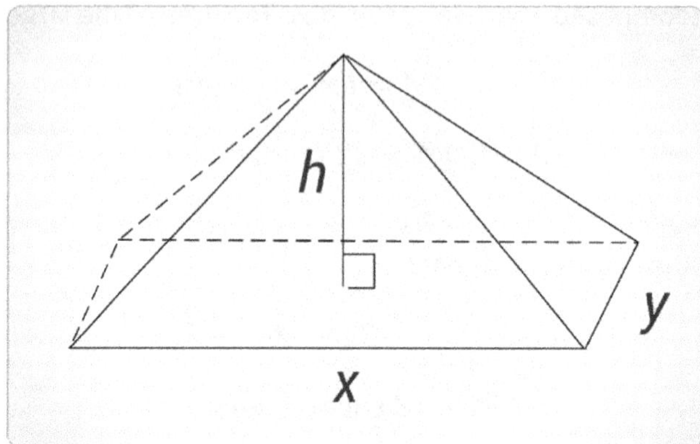

To find the surface area, the dimensions of each triangle need to be known. However, these dimensions can differ depending on the problem in question. Therefore, there is no general formula for calculating total surface area.

A **sphere** is a set of points all of which are equidistant from some central point. It is like a circle, but in three dimensions. The volume of a sphere of radius *r* is given by:

$$V = \frac{4}{3}\pi r^3$$

The surface area is given by:

$$A = 4\pi r^2$$

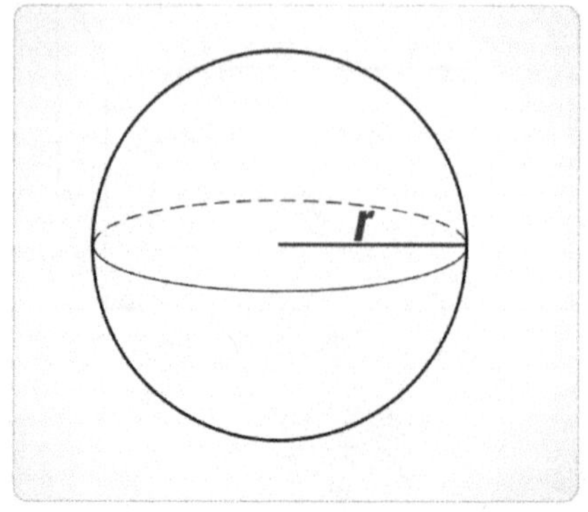

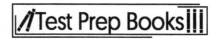

Lines and Angles

In geometry, a **line** connects two points, has no thickness, and extends indefinitely in both directions beyond the points. If it does end at two points, it is known as a **line segment**. It is important to distinguish between a line and a line segment.

An **angle** can be visualized as a corner. It is defined as the formation of two rays connecting at a vertex that extend indefinitely. Angles are usually measured in degrees. Their measurement is a measure of rotation. A full rotation equals 360 degrees and represents a circle. Half of a rotation equals 180 degrees and represents a half-circle. Subsequently, 90 degrees represents a quarter-circle. Similar to the hands on a clock, an angle begins at the center point, and two lines extend indefinitely from that point in two different directions.

A clock can be useful when learning how to measure angles. At 3:00, the big hand is on the 12 and the small hand is on the 3. The angle formed is 90 degrees and is known as a **right angle**. Any angle less than 90 degrees, such as the one formed at 2:00, is known as an **acute angle**. Any angle greater than 90 degrees is known as an **obtuse angle**. The entire clock represents 360 degrees, and each clockwise increment on the clock represents an addition of 30 degrees. Therefore, 6:00 represents 180 degrees, 7:00 represents 210 degrees, etc. Angle measurement is additive. An angle can be broken into two non-overlapping angles. The total measure of the larger angle is equal to the sum of the measurements of the two smaller angles.

A **ray** is a straight path that has an endpoint on one end and extends indefinitely in the other direction. Lines are known as being **coplanar** if they are located in the same plane. Coplanar lines exist within the same two-dimensional surface. Two lines are **parallel** if they are coplanar, extend in the same direction, and never cross. They are known as being **equidistant** because they are always the same distance from each other. If lines do cross, they are known as **intersecting lines**. As discussed previously, angles are utilized throughout geometry, and their measurement can be seen through the use of an analog clock. An angle is formed when two rays begin at the same endpoint. **Adjacent angles** can be formed by forming two angles out of one shared ray. They are two side-by-side angles that also share an endpoint.

Perpendicular lines are coplanar lines that form a right angle at their point of intersection. A triangle that contains a right angle is known as a **right triangle**. The sum of the angles within any triangle is always 180 degrees. Therefore, in a right triangle, the sum of the two angles that are not right angles is 90 degrees. Any two angles that sum up to 90 degrees are known as **complementary angles**. A triangle that contains an obtuse angle is known as an **obtuse triangle**. A triangle that contains three acute angles is known as an **acute triangle**. Here is an example of a 180-degree angle, split up into an acute and obtuse angle:

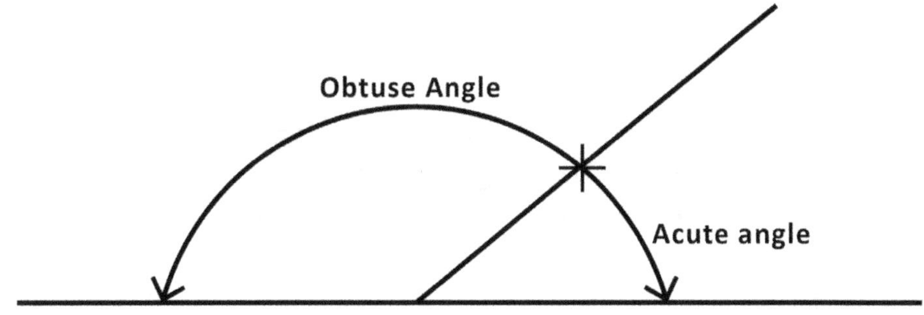

Relationships between Angles

Supplementary angles add up to 180 degrees. **Vertical angles** are two nonadjacent angles formed by two intersecting lines. For example, in the following picture, angles 4 and 2 are vertical angles and so are angles 1 and 3:

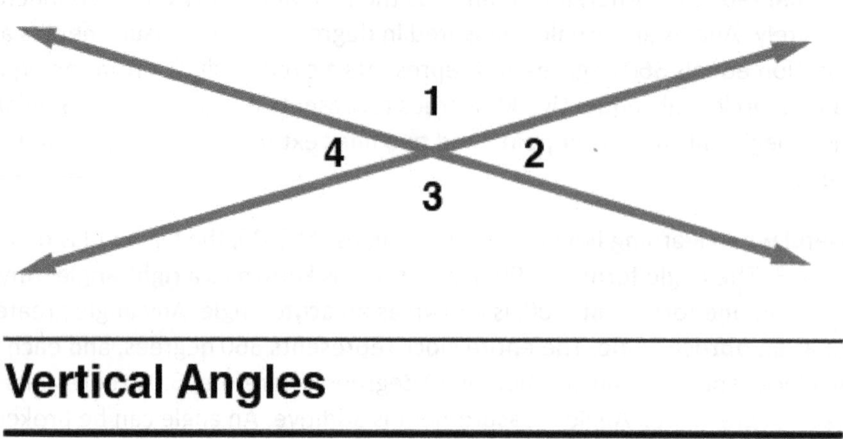

Angles that add up to 90 degrees are **complementary. Corresponding angles** are two angles in the same position whenever a straight line (known as a **transversal**) crosses two others. If the two lines are parallel, the corresponding angles are equal. In the following diagram, angles 1 and 3 are corresponding angles but aren't equal to each other:

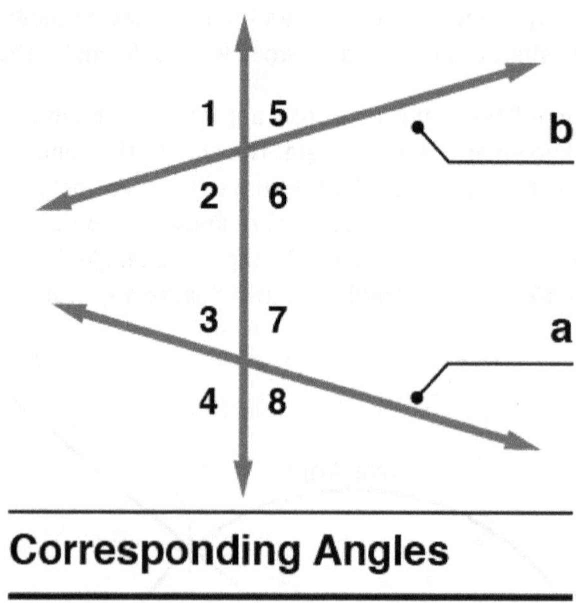

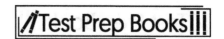

Alternate interior angles are also a pair of angles formed when two lines are crossed by a transversal. They are opposite angles that exist inside of the two lines. In the corresponding angles diagram above, angles 2 and 7 are alternate interior angles, as well as angles 6 and 3. **Alternate exterior angles** are opposite angles formed by a transversal but, in contrast to interior angles, exterior angles exist outside the two original lines. Therefore, angles 1 and 8 are alternate exterior angles and so are angles 5 and 4. Finally, **consecutive interior angles** are pairs of angles formed by a transversal. These angles are located on the same side of the transversal and inside the two original lines. Therefore, angles 2 and 3 are a pair of consecutive interior angles, and so are angles 6 and 7. These definitions are instrumental in solving many problems that involve determining relationships between angles. For example, the following problem utilizes the definition of complementary angles.

Two angles are complementary. If one angle is four times the other angle, what is the measure of each angle?

The first step is to determine the unknown, which is the measure of the angle.

The second step is to translate the problem into the equation using the known statement: the sum of two complementary angles is 90°. The resulting equation is $4x + x = 90$. The equation can be solved as follows:

$5x = 90$	Combine like terms on the left side of the equation
$x = 18$	Divide both sides of the equation by 5

The first angle is 18° and the second angle is 4 times the unknown, which is 4 times 18, or 72°.

Similarity and Congruence

Two figures are **congruent** if they have the same shape and same size. The two figures could have been rotated, reflected, or translated. Two figures are similar if they have been rotated, reflected, translated, and resized. Angle measure is preserved in similar figures. Both angle and side length are preserved in congruent figures.

In **similar figures**, if the ratio of two corresponding sides is known, then that ratio—or **scale factor**—holds true for all of the dimensions of the new figure.

Here is an example of applying this principle. Suppose that Lara is 5 feet tall and is standing 30 feet from the base of a light pole, and her shadow is 6 feet long. How high is the light on the pole? To figure this, it helps to make a sketch of the situation:

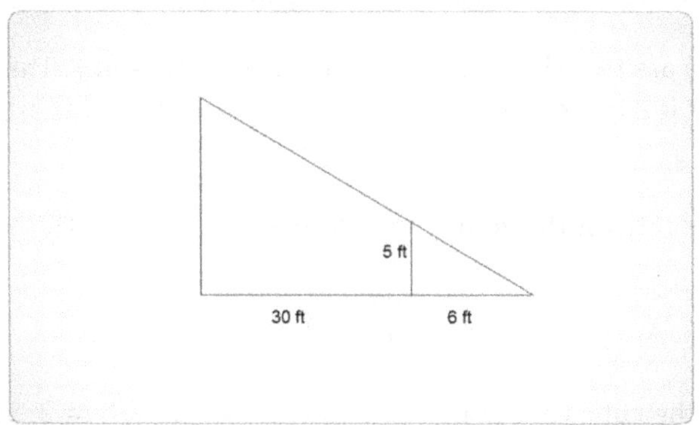

The light pole is the left side of the triangle. Lara is the 5-foot vertical line. Notice that there are two right triangles here, and that they have all the same angles as one another. Therefore, they form similar triangles and there is a ratio of proportionality between them.

The bases of these triangles are known. The small triangle, formed by Lara and her shadow, has a base of 6 feet. The large triangle, formed by the light pole along with the line from the base of the pole out to the end of Lara's shadow is $30 + 6 = 36$ feet long. So, the ratio of the big triangle to the little triangle will be $\frac{36}{6} = 6$. The height of the little triangle is 5 feet. Therefore, the height of the big triangle will be $6 \times 5 = 30$ feet, meaning that the light is 30 feet up the pole.

Notice that the perimeter of a figure changes by the ratio of proportionality between two similar figures, but the area changes by the square of the ratio. This is because if the length of one side is doubled, the area is quadrupled.

As an example, suppose two rectangles are similar, but the edges of the second rectangle are three times longer than the edges of the first rectangle. The area of the first rectangle is 10 square inches. How much more area does the second rectangle have than the first?

To answer this, note that the area of the second rectangle is $3^2 = 9$ times the area of the first rectangle, which is 10 square inches. Therefore, the area of the second rectangle is going to be $9 \times 10 = 90$ square inches. This means it has $90 - 10 = 80$ square inches more area than the first rectangle.

As a second example, suppose *X* and *Y* are similar right triangles. The hypotenuse of *X* is 4 inches. The area of *Y* is $\frac{1}{4}$ the area of *X*. What is the hypotenuse of *Y*?

First, realize the area has changed by a factor of $\frac{1}{4}$. The area changes by a factor that is the *square* of the ratio of changes in lengths, so the ratio of the lengths is the square root of the ratio of areas. That means that the ratio of lengths must be is:

$$\sqrt{\frac{1}{4}} = \frac{1}{2}$$

and the hypotenuse of *Y* must be $\frac{1}{2} \times 4 = 2$ inches.

Volumes between similar solids change like the cube of the change in the lengths of their edges. Likewise, if the ratio of the volumes between similar solids is known, the ratio between their lengths is calculated by finding the cube root of the ratio of their volumes.

For example, suppose there are two similar rectangular pyramids *X* and *Y*. The base of *X* is 1 inch by 2 inches, and the volume of *X* is 8 inches. The volume of *Y* is 64 inches. What are the dimensions of the base of *Y*?

To answer this, first find the ratio of the volume of *Y* to the volume of *X*. This will be given by:

$$\frac{64}{8} = 8$$

Now the ratio of lengths is the cube root of the ratio of volumes, or $\sqrt[3]{8} = 2$. So, the dimensions of the base of *Y* must be 2 inches by 4 inches.

The criteria needed to prove triangles are congruent involves both angle and side congruence. Both pairs of related angles and sides need to be of the same measurement to use congruence in a proof. The criteria to prove similarity in triangles involves proportionality of side lengths. Angles must be congruent in similar triangles; however, corresponding side lengths only need to be a constant multiple of each other. Once similarity is established, it can be used in proofs as well. Relationships in geometric figures other than triangles can be proven using triangle congruence and similarity. If a similar or congruent triangle can be found within another type of geometric figure, their criteria can be used to prove a relationship about a given formula. For instance, a rectangle can be broken up into two congruent triangles.

If two angles of one triangle are congruent with two angles of a second triangle, the triangles are similar. This is because, within any triangle, the sum of the angle measurements is 180 degrees. Therefore, if two are congruent, the third angle must also be congruent because their measurements are equal. Three congruent pairs of angles mean that the triangles are similar.

There are five theorems to show that triangles are congruent when it's unknown whether each pair of angles and sides are congruent. Each theorem is a shortcut that involves different combinations of sides and angles that must be the same for the two triangles to be congruent. For example, **side-side-side (SSS)** states that if all sides are equal, the triangles are congruent. **Side-angle-side (SAS)** states that if two pairs of sides are equal and the included angles are congruent, then the triangles are congruent. Similarly, **angle-side-angle (ASA)** states that if two pairs of angles are congruent and the included side lengths are equal, the triangles are similar.

Angle-angle-side (AAS) states that two triangles are congruent if they have two pairs of congruent angles and a pair of corresponding equal side lengths that aren't included. Finally, **hypotenuse-leg (HL)** states that if two right triangles have equal hypotenuses and an equal pair of shorter sides, then the triangles are congruent. An important item to note is that angle-angle-angle *(AAA)* is not enough information to have congruence. It's important to understand why these rules work by using rigid motions to show congruence between the triangles with the given properties. For example, three reflections are needed to show why *SAS* follows from the definition of congruence.

Measuring Lengths of Objects

The length of an object can be measured using standard tools such as rulers, yard sticks, meter sticks, and measuring tapes. The following image depicts a yardstick:

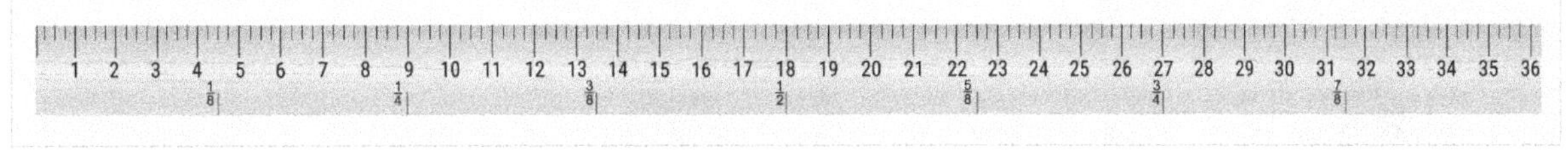

Choosing the right tool to perform the measurement requires determining whether United States customary units or metric units are desired, and having a grasp of the approximate length of each unit and the approximate length of each tool. The measurement can still be performed by trial and error without the knowledge of the approximate size of the tool.

For example, if you were asked to determine the length of a room in feet, a United States customary unit, you could theoretically use a few different tools for this task. These include a ruler (typically 12 inches/1 foot long), a yardstick (3 feet/1 yard long), or a tape measure displaying feet (typically either 25 feet or 50 feet). Because the length of a room is much larger than the length of a ruler or a yardstick, a tape measure should be used to perform the measurement.

When the correct measuring tool is selected, the measurement is performed by first placing the tool directly above or below the object (if making a horizontal measurement) or directly next to the object (if making a vertical measurement). The next step is aligning the tool so that one end of the object is at the mark for zero units, then recording the unit of the mark at the other end of the object. To give the length of a paperclip in metric units, a ruler displaying centimeters is aligned with one end of the paper clip to the mark for zero centimeters.

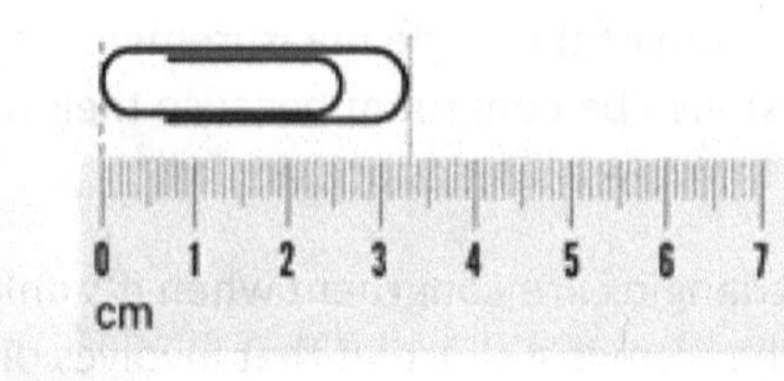

Directly down from the other end of the paperclip is the mark that measures its length. In this case, that mark is two small dashes past the 3 centimeter mark. Each small dash is 1 millimeter (or .1 centimeters). Therefore, the length of the paper clip is 3.2 centimeters.

To compare the lengths of objects, each length must be expressed with the same units. If possible, the objects should be measured with the same tool or with tools utilizing the same units.

For example, a ruler and a yardstick can both measure length in inches. If the lengths of the objects are expressed in different units, these different units must be converted to the same unit before comparing them. If two lengths are expressed in the same unit, the lengths may be compared by subtracting the smaller value from the larger value. For example, suppose the lengths of two gardens are to be compared.

Garden A has a length of 4 feet, and garden B has a length of 2 yards. 2 yards is converted to 6 feet so that the measurements have similar units. Then, the smaller length (4 feet) is subtracted from the larger length (6ft): 6ft – 4ft = 2ft. Therefore, garden B is 2 feet larger than garden A.

Relative Sizes of United States Customary Units and Metric Units

The United States customary system and the metric system each consist of distinct units to measure lengths and volume of liquids. The U.S. customary units for length, from smallest to largest, are: inch (in), foot (ft), yard (yd), and mile (mi). The metric units for length, from smallest to largest, are: millimeter (mm), centimeter (cm), decimeter (dm), meter (m), and kilometer (km). The relative size of each unit of length is shown below.

U.S. Customary	Metric	Conversion
12in = 1ft	10mm = 1cm	1in = 2.54cm
36in = 3ft = 1yd	10cm = 1dm(decimeter)	1m ≈ 3.28ft ≈ 1.09yd
5,280ft = 1,760yd = 1mi	100cm = 10dm = 1m	1mi ≈ 1.6km
	1000m = 1km	

The U.S. customary units for volume of liquids, from smallest to largest, are: fluid ounces (fl oz), cup (c), pint (pt), quart (qt), and gallon (gal). The metric units for volume of liquids, from smallest to largest, are: milliliter (mL), centiliter (cL), deciliter (dL), liter (L), and kiloliter (kL).

The relative size of each unit of liquid volume is shown below:

U.S. Customary	Metric	Conversion
8fl oz = 1c	10mL = 1cL	1pt ≈ 0.473L
2c = 1pt	10cL = 1dL	1L ≈ 1.057qt
4c = 2pt = 1qt	1,000mL = 100cL = 10dL = 1L	1gal ≈ 3.785L
4qt = 1gal	1,000L = 1kL	

The U.S. customary system measures weight (how strongly Earth is pulling on an object) in the following units, from least to greatest: ounce (oz), pound (lb), and ton. The metric system measures mass (the quantity of matter within an object) in the following units, from least to greatest: milligram (mg), centigram (cg), gram (g), kilogram (kg), and metric ton (MT). The relative sizes of each unit of weight and mass are shown below.

U.S. Measures of Weight	Metric Measures of Mass
16oz = 1lb	10mg = 1cg
2,000lb = 1 ton	100cg = 1g
	1,000g = 1kg
	1,000kg = 1MT

Note that weight and mass DO NOT measure the same thing.

Time is measured in the following units, from shortest to longest: second (sec), minute (min), hour (h), day (d), week (wk), month (mo), year (yr), decade, century, millennium. The relative sizes of each unit of time is shown below.

- 60sec = 1min
- 60min = 1h
- 24hr = 1d
- 7d = 1wk

- 52wk = 1yr
- 12mo = 1yr
- 10yr = 1 decade
- 100yrs = 1 century
- 1,000yrs = 1 millennium

Conversion of Units

When working with different systems of measurement, conversion from one unit to another may be necessary. The conversion rate must be known to convert units. One method for converting units is to write and solve a proportion. The arrangement of values in a proportion is extremely important. Suppose that a problem requires converting 20 fluid ounces to cups. To do so, a proportion can be written using the conversion rate of 8fl oz = 1c with *x* representing the missing value.

The proportion can be written in any of the following ways:

$$\frac{1}{8} = \frac{x}{20}\left(\frac{c\ for\ conversion}{fl\ oz\ for\ conversion} = \frac{unknown\ c}{fl\ oz\ given}\right)$$

$$\frac{8}{1} = \frac{20}{x}\left(\frac{fl\ oz\ for\ conversion}{c\ for\ conversion} = \frac{fl\ oz\ given}{unknown\ c}\right)$$

$$\frac{1}{x} = \frac{8}{20}\left(\frac{c\ for\ conversion}{unknown\ c} = \frac{fl\ oz\ for\ conversion}{fl\ oz\ given}\right)$$

$$\frac{x}{1} = \frac{20}{8}\left(\frac{unknown\ c}{c\ for\ conversion} = \frac{fl\ oz\ given}{fl\ oz\ for\ conversion}\right)$$

To solve a proportion, the ratios are cross-multiplied and the resulting equation is solved. When cross-multiplying, all four proportions above will produce the same equation:

$$(8)(x) = (20)(1) \rightarrow 8x = 20$$

Dividing by 8 to isolate the variable *x*, the result is *x* = 2.5. The variable *x* represented the unknown number of cups. Therefore, the conclusion is that 20 fluid ounces converts (is equal) to 2.5 cups.

Sometimes converting units requires writing and solving more than one proportion. Suppose an exam question asks to determine how many hours are in 2 weeks. Without knowing the conversion rate between hours and weeks, this can be determined knowing the conversion rates between weeks and days, and between days and hours. First, weeks are converted to days, then days are converted to hours. To convert from weeks to days, the following proportion can be written:

$$\frac{7}{1} = \frac{x}{2}\left(\frac{days\ conversion}{weeks\ conversion} = \frac{days\ unknown}{weeks\ given}\right)$$

Cross-multiplying produces:

$$(7)(2) = (x)(1)$$

$$14 = x$$

Therefore, 2 weeks is equal to 14 days. Next, a proportion is written to convert 14 days to hours:

$$\frac{24}{1}=\frac{x}{14}\left(\frac{conversion\ hours}{conversion\ days}=\frac{unknown\ hours}{given\ days}\right)$$

Cross-multiplying produces:

$$(24)(14)=(x)(1)\rightarrow 336=x$$

Therefore, the answer is that there are 336 hours in 2 weeks.

Data Analysis/Probability

Graphical Representation of Data

Various graphs can be used to visually represent a given set of data. Each type of graph requires a different method of arranging data points and different calculations of the data. To construct a **histogram**, the range of the data points is divided into equal intervals. The frequency for each interval is then determined, which reveals how many points fall into each interval. A **graph** is constructed with the vertical axis representing the frequency and the horizontal axis representing the intervals. The lower value of each interval should be labeled along the horizontal axis. Finally, for each interval, a bar is drawn from the lower value of each interval to the lower value of the next interval with a height equal to the frequency of the interval. Because of the intervals, histograms do not have any gaps between bars along the horizontal axis.

To construct a **box** (or **box-and-whisker**) **plot**, the five number summary for the data set is calculated as follows: the second quartile (Q_2) is the median of the set. The first quartile (Q_1) is the median of the values below Q_2. The third quartile (Q_3) is the median of the values above Q_2. The upper extreme is the highest value in the data set if it is not an outlier (greater than 1.5 times the interquartile range Q_3 - Q_1). The lower extreme is the least value in the data set if it is not an outlier (more than 1.5 times lower than the interquartile range). To construct the box-and-whisker plot, each value is plotted on a number line, along with any outliers. The **box** consists of Q_1 and Q_3 as its top and bottom and Q_2 as the dividing line inside the box. The **whiskers** extend from the lower extreme to Q_1 and from Q_3 to the upper extreme.

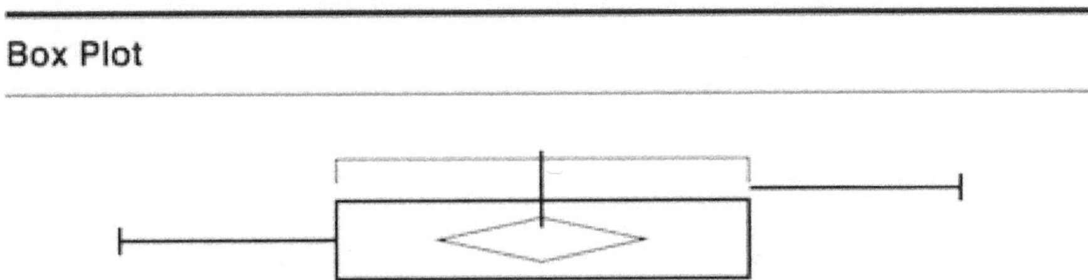

A scatter plot displays the relationship between two variables. Values for the independent variable, typically denoted by *x*, are paired with values for the dependent variable, typically denoted by *y*. Each set of corresponding values are written as an ordered pair (*x*, *y*). To construct the graph, a coordinate grid is labeled with the *x*-axis representing the independent variable and the *y*-axis representing the dependent variable. Each ordered pair is graphed.

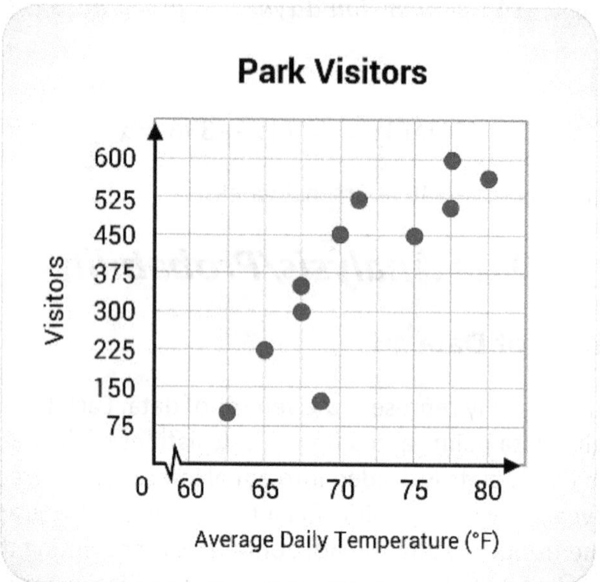

Like a scatter plot, a **line graph** compares variables that change continuously, typically over time. Paired data values (ordered pair) are plotted on a coordinate grid with the *x*- and *y*-axis representing the variables. A line is drawn from each point to the next, going from left to right. The line graph below displays cell phone use for given years (two variables) for men, women, and both sexes (three data sets).

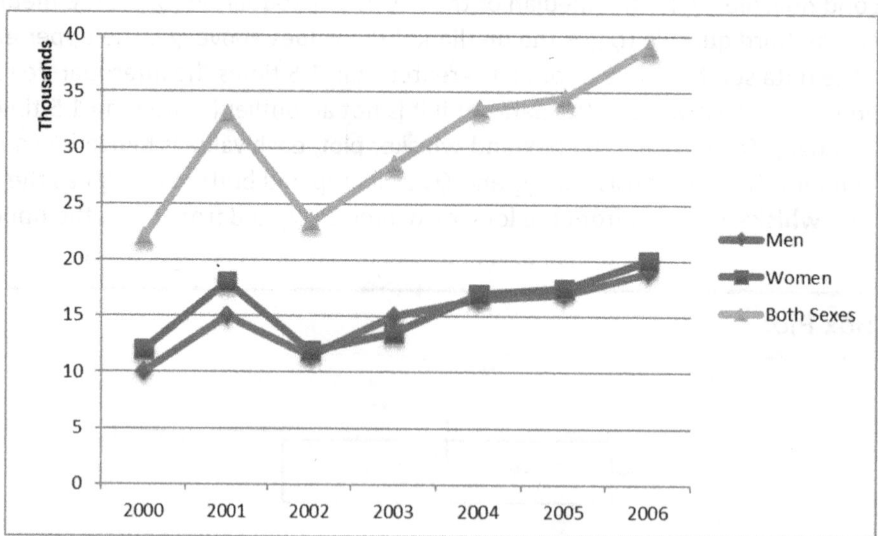

A **line plot**, also called **dot plot**, displays the frequency of data (numerical values) on a number line. To construct a line plot, a number line is used that includes all unique data values. It is marked with x's or dots above the value the number of times that the value occurs in the data set.

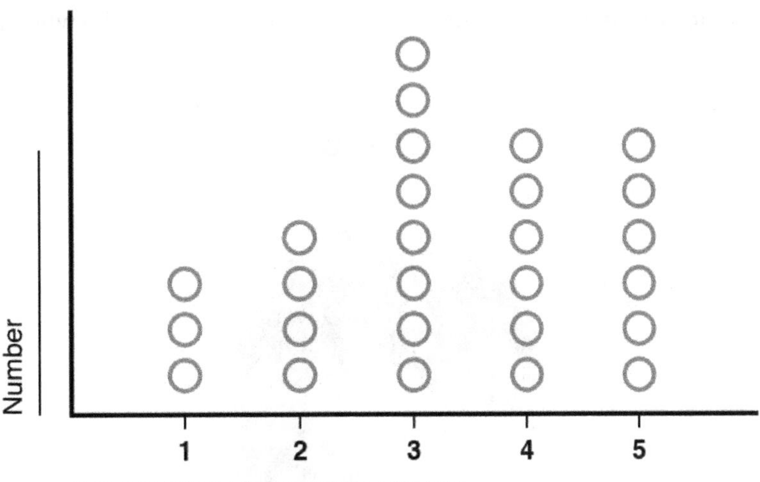

A **bar graph** is a diagram in which the quantity of items within a specific classification is represented by the height of a rectangle. Each type of classification is represented by a rectangle of equal width.

Here is an example of a bar graph:

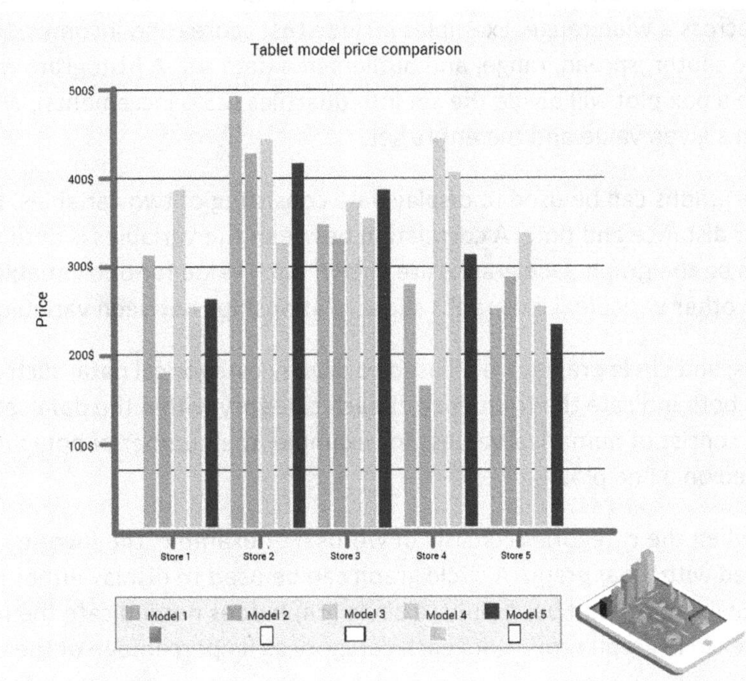

A **circle graph**, also called a **pie chart**, shows categorical data with each category representing a percentage of the whole data set. To make a circle graph, the percent of the data set for each category must be determined. To do so, the frequency of the category is divided by the total number of data points and converted to a percent. For example, if 80 people were asked what their favorite sport is and 20 responded basketball, basketball makes up 25% of the data ($\frac{20}{80}$ =.25=25%). Each category in a data set is represented by a slice of the circle proportionate to its percentage of the whole.

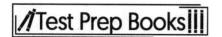

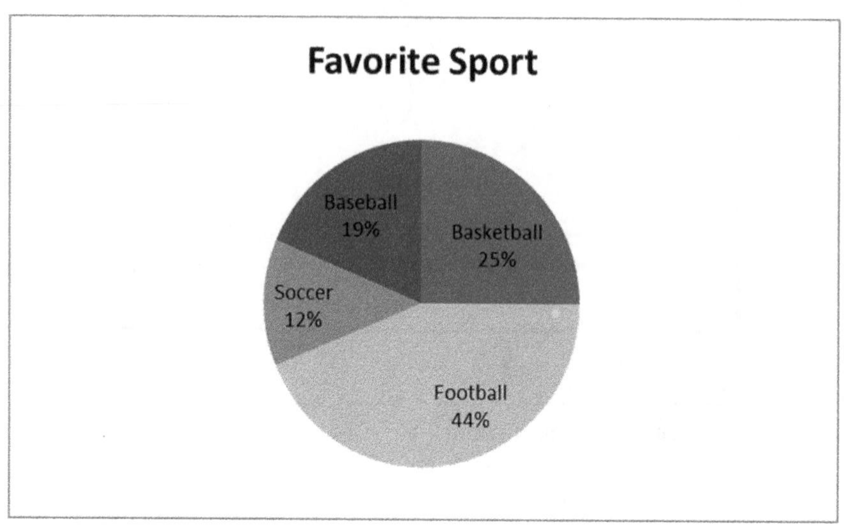

Choice of Graphs to Display Data

Choosing the appropriate graph to display a data set depends on what type of data is included in the set and what information must be displayed. Histograms and box plots can be used for data sets consisting of individual values across a wide range. Examples include test scores and incomes. Histograms and box plots will indicate the center, spread, range, and outliers of a data set. A histogram will show the shape of the data set, while a box plot will divide the set into quartiles (25% increments), allowing for comparison between a given value and the entire set.

Scatter plots and line graphs can be used to display data consisting of two variables. Examples include height and weight, or distance and time. A correlation between the variables is determined by examining the points on the graph. Line graphs are used if each value for one variable pairs with a distinct value for the other variable. Line graphs show relationships between variables.

Line plots, bar graphs, and circle graphs are all used to display categorical data, such as surveys. Line plots and bar graphs both indicate the frequency of each category within the data set. A line plot is used when the categories consist of numerical values. For example, the number of hours of TV watched by individuals is displayed on a line plot.

A bar graph is used when the categories consists of words. For example, the favorite ice cream of individuals is displayed with a bar graph. A circle graph can be used to display either type of categorical data. However, unlike line plots and bar graphs, a circle graph does not indicate the frequency of each category. Instead, the circle graph represents each category as its percentage of the whole data set.

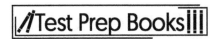

Measures of Center and Range

The center of a set of data (statistical values) can be represented by its mean, median, or mode. These are sometimes referred to as measures of central tendency. The **mean** is the average of the data set. The mean can be calculated by adding the data values and dividing by the sample size (the number of data points). Suppose a student has test scores of 93, 84, 88, 72, 91, and 77. To find the mean, or average, the scores are added and the sum is divided by 6 because there are 6 test scores:

$$\frac{93+84+88+72+91+77}{6}=\frac{505}{6}=84.17$$

Given the mean of a data set and the sum of the data points, the sample size can be determined by dividing the sum by the mean. Suppose you are told that Kate averaged 12 points per game and scored a total of 156 points for the season. The number of games that she played (the sample size or the number of data points) can be determined by dividing the total points (sum of data points) by her average (mean of data points): $\frac{156}{12}=13$. Therefore, Kate played in 13 games this season.

If given the mean of a data set and the sample size, the sum of the data points can be determined by multiplying the mean and sample size. Suppose you are told that Tom worked 6 days last week for an average of 5.5 hours per day. The total number of hours worked for the week (sum of data points) can be determined by multiplying his daily average (mean of data points) by the number of days worked (sample size): $5.5 \times 6 = 33$. Therefore, Tom worked a total of 33 hours last week.

The **median** of a data set is the value of the data point in the middle when the sample is arranged in numerical order. To find the median of a data set, the values are written in order from least to greatest. The lowest and highest values are simultaneously eliminated, repeating until the value in the middle remains. Suppose the salaries of math teachers are:

$35,000; $38,500; $41,000; $42,000; $42,000; $44,500; $49,000

The values are listed from least to greatest to find the median. The lowest and highest values are eliminated until only the middle value remains. Repeating this step three times reveals a median salary of $42,000. If the sample set has an even number of data points, two values will remain after all others are eliminated. In this case, the mean of the two middle values is the median. Consider the following data set: 7, 9, 10, 13, 14, 14. Eliminating the lowest and highest values twice leaves two values, 10 and 13, in the middle. The mean of these values $\left(\frac{10+13}{2}\right)$ is the median. Therefore, the set has a median of 11.5.

The **mode** of a data set is the value that appears most often. A data set may have a single mode, multiple modes, or no mode. If different values repeat equally as often, multiple modes exist. If no value repeats, no mode exists. Consider the following data sets:

- A: 7, 9, 10, 13, 14, 14
- B: 37, 44, 33, 37, 49, 44, 51, 34, 37, 33, 44
- C: 173, 154, 151, 168, 155

Set A has a mode of 14. Set B has modes of 37 and 44. Set C has no mode.

The **range** of a data set is the difference between the highest and the lowest values in the set. The range can be considered to be the span of the data set. To determine the range, the smallest value in the set is subtracted from the largest value. The ranges for the data sets A, B, and C above are calculated as follows:

$$A: 14 - 7 = 7$$
$$B: 51 - 33 = 18$$
$$C: 173 - 151 = 22$$

Best Description of a Set of Data

Measures of central tendency, namely mean, median, and mode, describe characteristics of a set of data. Specifically, they are intended to represent a *typical* value in the set by identifying a central position of the set. Depending on the characteristics of a specific set of data, different measures of central tendency are more indicative of a typical value in the set.

When a data set is grouped closely together with a relatively small range and the data is spread out somewhat evenly, the mean is an effective indicator of a typical value in the set. Consider the following data set representing the height of sixth grade boys in inches: 61 inches, 54 inches, 58 inches, 63 inches, 58 inches. The mean of the set is 58.8 inches. The data set is grouped closely (the range is only 9 inches) and the values are spread relatively evenly (three values below the mean and two values above the mean). Therefore, the mean value of 58.8 inches is an effective measure of central tendency in this case.

When a data set contains a small number of values, with one either extremely large or extremely small when compared to the other values, the mean is not an effective measure of central tendency. Consider the following data set representing annual incomes of homeowners on a given street: $71,000; $74,000; $75,000; $77,000; $340,000. The mean of this set is $127,400. This figure does not indicate a typical value in the set, which contains four out of five values between $71,000 and $77,000. The median is a much more effective measure of central tendency for data sets such as these. Finding the middle value diminishes the influence of outliers, or numbers that may appear out of place, like the $340,000 annual income. The median for this set is $75,000 which is much more typical of a value in the set.

The mode of a data set is a useful measure of central tendency for categorical data when each piece of data is an option from a category. Consider a survey of 31 commuters asking how they get to work with results summarized below.

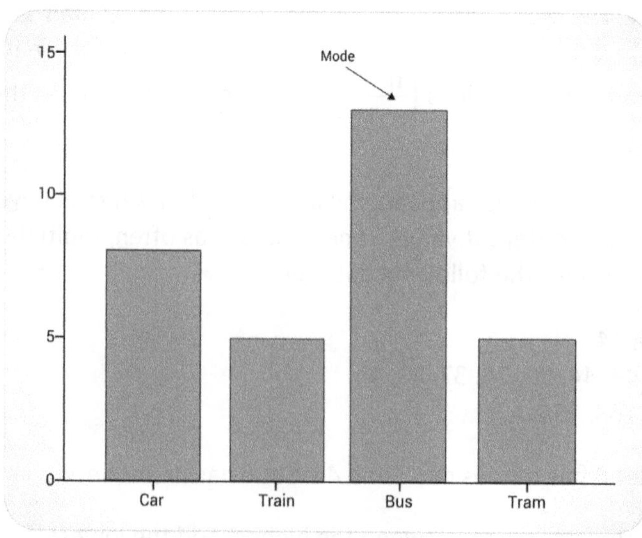

The mode for this set represents the value, or option, of the data that repeats most often. This indicates that the bus is the most popular method of transportation for the commuters.

Effects of Changes in Data

Changing all values of a data set in a consistent way produces predictable changes in the measures of the center and range of the set. A linear transformation changes the original value into the new value by either adding a given number to each value, multiplying each value by a given number, or both. Adding (or subtracting) a given value to each data point will increase (or decrease) the mean, median, and any modes by the same value. However, the range will remain the same due to the way that range is calculated. Multiplying (or dividing) a given value by each data point will increase (or decrease) the mean, median, and any modes, and the range by the same factor.

Consider the following data set, call it set *P*, representing the price of different cases of soda at a grocery store: $4.25, $4.40, $4.75, $4.95, $4.95, $5.15. The mean of set *P* is $4.74. The median is $4.85. The mode of the set is $4.95. The range is $0.90. Suppose the state passes a new tax of $0.25 on every case of soda sold. The new data set, set *T*, is calculated by adding $0.25 to each data point from set *P*. Therefore, set *T* consists of the following values: $4.50, $4.65, $5.00, $5.20, $5.20, $5.40. The mean of set *T* is $4.99. The median is $5.10. The mode of the set is $5.20. The range is $.90. The mean, median and mode of set *T* is equal to $0.25 added to the mean, median, and mode of set *P*. The range stays the same.

Now suppose, due to inflation, the store raises the cost of every item by 10 percent. Raising costs by 10 percent is calculated by multiplying each value by 1.1. The new data set, set *I*, is calculated by multiplying each data point from set *T* by 1.1. Therefore, set *I* consists of the following values: $4.95, $5.12, $5.50, $5.72, $5.72, $5.94. The mean of set *I* is $5.49. The median is $5.61. The mode of the set is $5.72. The range is $0.99. The mean, median, mode, and range of set *I* is equal to 1.1 multiplied by the mean, median, mode, and range of set *T* because each increased by a factor of 10 percent.

Describing a Set of Data

A set of data can be described in terms of its center, spread, shape and any unusual features. The center of a data set can be measured by its mean, median, or mode. The spread of a data set refers to how far the data points are from the center (mean or median). The spread can be measured by the range or the quartiles and interquartile range. A data set with data points clustered around the center will have a small spread. A data set covering a wide range will have a large spread.

When a data set is displayed as a **histogram** or frequency distribution plot, the shape indicates if a sample is normally distributed, symmetrical, or has measures of skewness or kurtosis. When graphed, a data set with a **normal distribution** will resemble a bell curve.

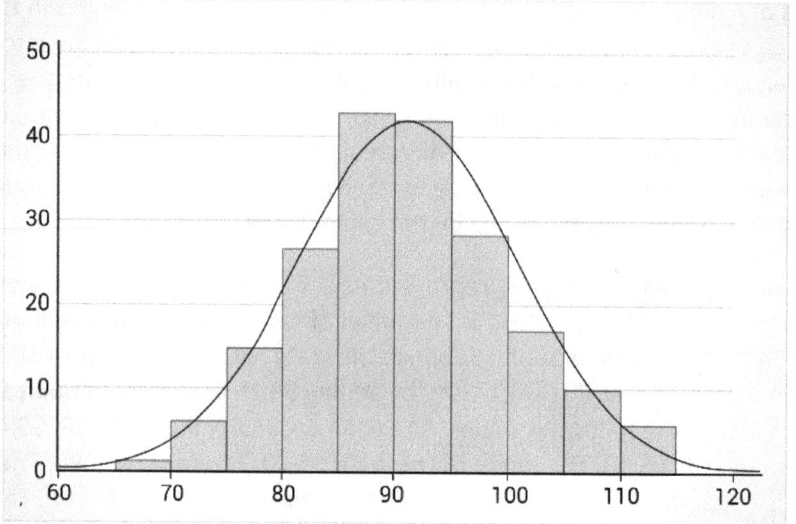

If the data set is symmetrical, each half of the graph when divided at the center is a mirror image of the other. If the graph has fewer data points to the right, the data is **skewed right**. If it has fewer data points to the left, the data is **skewed left**.

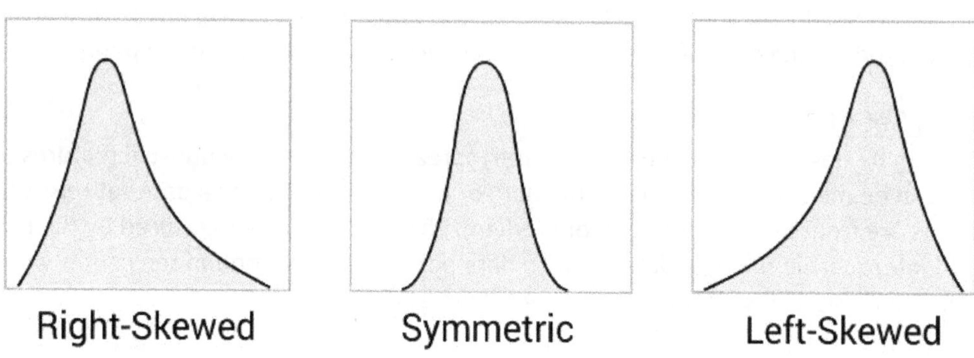

Kurtosis is a measure of whether the data is heavy-tailed with a high number of outliers, or light-tailed with a low number of outliers.

A description of a data set should include any unusual features such as gaps or outliers. A **gap** is a span within the range of the data set containing no data points. An **outlier** is a data point with a value either extremely large or extremely small when compared to the other values in the set.

Interpreting Displays of Data

A set of data can be visually displayed in various forms allowing for quick identification of characteristics of the set. Histograms, such as the one shown below, display the number of data points (vertical axis) that fall into given intervals (horizontal axis) across the range of the set. Suppose the histogram below displays IQ scores of students. Histograms can display the center, spread, shape, and any unusual characteristics of a data set.

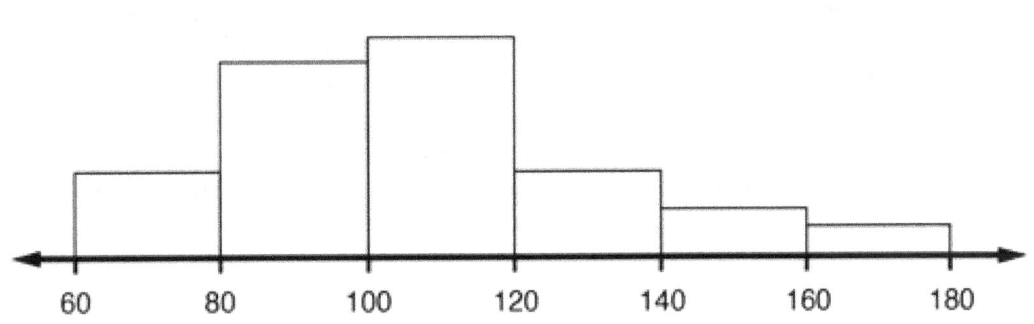

As mentioned, a box plot, also called a box-and-whisker plot, divides the data points into four groups and displays the five-number summary for the set, as well as any outliers. The five-number summary consists of:

- The lower extreme: the lowest value that is not an outlier
- The higher extreme: the highest value that is not an outlier
- The median of the set: also referred to as the second quartile or Q_2
- The first quartile or Q_1: the median of values below Q_2
- The third quartile or Q_3: the median of values above Q_2

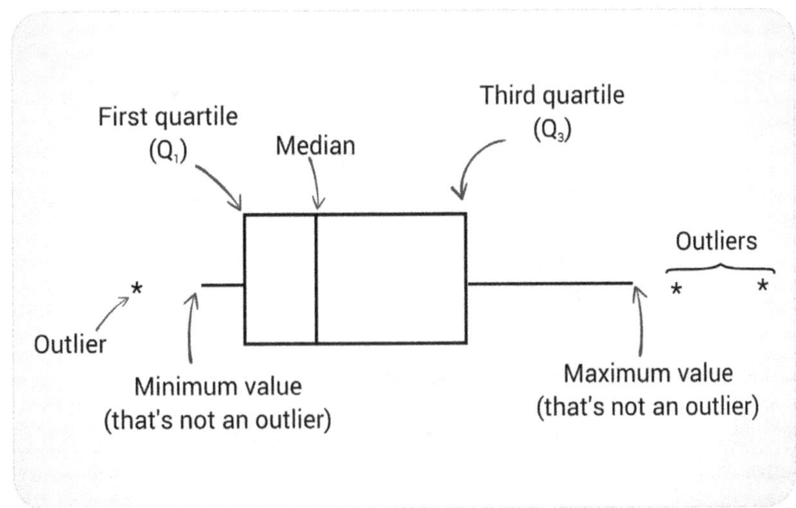

Suppose the box plot displays IQ scores for 12th grade students at a given school. The five number summary of the data consists of: lower extreme (67); upper extreme (127); Q_2or median (100); Q_1 (91); Q_3 (108); and outliers (135 and 140). Although all data points are not known from the plot, the points are divided into four quartiles each, including 25% of the data points. Therefore, 25% of students scored between 67 and 91, 25% scored between 91 and 100, 25% scored between 100 and 108, and 25% scored between 108 and 127. These percentages include the normal values for the set and exclude the outliers. This information is useful when comparing a given score with the rest of the scores in the set.

A scatter plot is a mathematical diagram that visually displays the relationship or connection between two variables. The independent variable is placed on the *x*-axis (the horizontal axis), and the dependent variable is placed on the *y*-axis (the vertical axis). When visually examining the points on the graph, if the points model a linear relationship, or a line of best-fit can be drawn through the points with the points relatively close on either side, then a correlation exists. If the line of best-fit has a positive slope (rises from left to right), then the variables have a positive correlation. If, like the image below, the line of best-fit has a negative slope (falls from left to right), then the variables have a negative correlation. If a line of best-fit cannot be drawn, then no correlation exists. A positive or negative correlation can be categorized as strong or weak, depending on how closely the points are graphed around the line of best-fit.

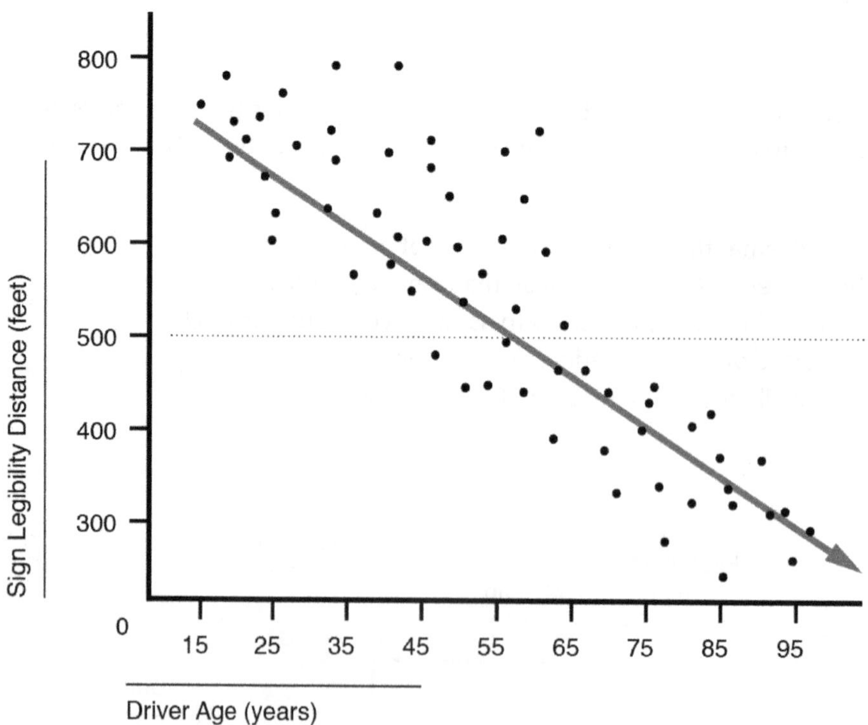

Probabilities Relative to Likelihood of Occurrence

Probability is a measure of how likely an event is to occur. Probability is written as a fraction or decimal between zero and one. If an event has a probability of zero, the event will never occur. If an event has a probability of one, the event will definitely occur. If the probability of an event is closer to zero, the event is unlikely to occur. If the probability of an event is closer to one, the event is more likely to occur.

For example, a probability of $\frac{1}{2}$ means that the event is equally as likely to occur as it is to not occur. An example of this is tossing a coin. To calculate the probability of an event, the number of favorable outcomes is divided by the number of total outcomes. For example, suppose you have 2 raffle tickets out of 20 total tickets sold. The probability that you win the raffle is calculated:

$$\frac{number\ of\ favorable\ outcomes}{total\ numberof\ outcomes} = \frac{2}{20} = \frac{1}{10} \text{(always reduce fractions)}$$

Therefore, the probability of winning the raffle is $\frac{1}{10}$ or 0.1.

Chance is the measure of how likely an event is to occur, written as a percent. If an event will never occur, the event has a 0% chance. If an event will certainly occur, the event has a 100% chance. If an event will sometimes occur, the event has a chance somewhere between 0% and 100%. To calculate chance, probability is calculated and the fraction or decimal is converted to a percent.

The probability of multiple events occurring can be determined by multiplying the probability of each event. For example, suppose you flip a coin with heads and tails, and roll a six-sided dice numbered one through six. To find the probability that you will flip heads AND roll a two, the probability of each event is determined and those fractions are multiplied. The probability of flipping heads is $\frac{1}{2}\left(\frac{1\ side\ with\ heads}{2\ sides\ total}\right)$ and the probability of rolling a two is $\frac{1}{6}\left(\frac{1\ side\ with\ a\ 2}{6\ total\ sides}\right)$. The probability of flipping heads AND rolling a 2 is:

$$\frac{1}{2} \times \frac{1}{6} = \frac{1}{12}$$

The above scenario with flipping a coin and rolling a dice is an example of independent events. **Independent events** are circumstances in which the outcome of one event does not affect the outcome of the other event. Conversely, **dependent events** are ones in which the outcome of one event affects the outcome of the second event. Consider the following scenario: a bag contains 5 black marbles and 5 white marbles. What is the probability of picking 2 black marbles without replacing the marble after the first pick?

The probability of picking a black marble on the first pick is:

$$\frac{5}{10}\left(\frac{5\ black\ marbles}{10\ total\ marbles}\right)$$

Assuming that a black marble was picked, there are now 4 black marbles and 5 white marbles for the second pick. Therefore, the probability of picking a black marble on the second pick is:

$$\frac{4}{9}\left(\frac{4\ black\ marbles}{9\ total\ marbles}\right)$$

To find the probability of picking two black marbles, the probability of each is multiplied:

$$\frac{5}{10} \times \frac{4}{9} = \frac{20}{90} = \frac{2}{9}$$

Practice Questions

1. At the beginning of the day, Xavier has 20 apples. At lunch, he meets his sister Emma and gives her half of his apples. After lunch, he stops by his neighbor Jim's house and gives him 6 of his apples. He then uses ¾ of his remaining apples to make an apple pie for dessert at dinner. At the end of the day, how many apples does Xavier have left?

   a. 4
   b. 6
   c. 2
   d. 1
   e. 3

2. Shawna buys $2\frac{1}{2}$ gallons of paint. If she uses $\frac{1}{3}$ of it on the first day, how much does she have left?

   a. $1\frac{5}{6}$ gallons
   b. $1\frac{1}{2}$ gallons
   c. $1\frac{1}{3}$ gallons
   d. 2 gallons
   e. $1\frac{2}{3}$ gallons

3. If $\frac{5}{2} \div \frac{1}{3} = n$, then n is between which of the following?

   a. 5 and 7
   b. 7 and 9
   c. 9 and 11
   d. 3 and 5
   e. 11 and 13

4. The graph shows the position of a car over a 10-second time interval. Which of the following is the correct interpretation of the graph for the interval 1 to 3 seconds?

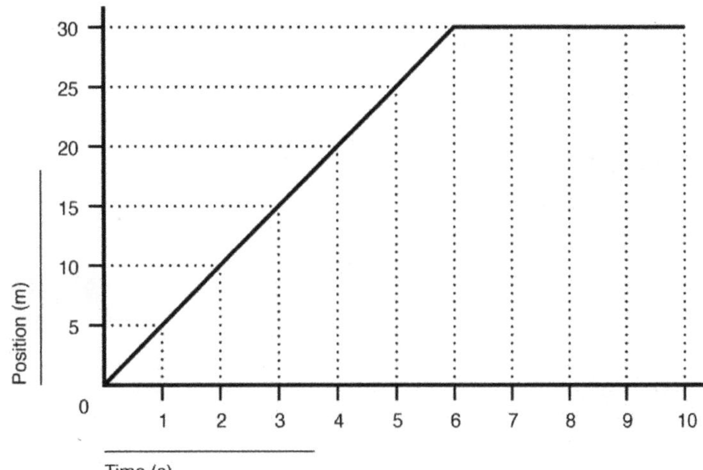

a. The car remains in the same position.
b. The car is traveling at a speed of 5m/s.
c. The car is traveling up a hill.
d. The car is traveling at 5mph.
e. The car accelerates at a rate of 5m/s.

5. There are $4x + 1$ treats in each party favor bag. If a total of $60x + 15$ treats are distributed, how many bags are given out?
a. 15
b. 16
c. 20
d. 21
e. 22

6. Apples cost \$2 each, while bananas cost \$3 each. Maria purchased 10 fruits in total and spent \$22. How many apples did she buy?
a. 4
b. 5
c. 6
d. 7
e. 8

7. What is the y-intercept of $y = x^{5/3} + (x - 3)(x + 1)$?
a. 3.5
b. 7.6
c. -3
d. -15.1
e. $\frac{5}{3}$

8. Dwayne has received the following scores on his math tests: 78, 92, 83, and 97. What score must Dwayne get on his next math test to have an overall average of 90?

a. 89
b. 98
c. 95
d. 94
e. 100

9. What are all the factors of 12?

a. 12, 24, 36
b. 1, 2, 4, 6, 12
c. 12, 24, 36, 48
d. 1, 2, 3, 4, 6, 12
e. 0, 1, 12

10. $3\frac{2}{3} - 1\frac{4}{5} =$

a. $1\frac{13}{15}$
b. $\frac{14}{15}$
c. $2\frac{2}{3}$
d. $\frac{4}{5}$
e. $\frac{4}{15}$

11. A rectangle has a length that is 5 feet longer than three times its width. If the perimeter is 90 feet, what is the length in feet?

a. 10
b. 20
c. 25
d. 30
e. 35

12. Five of six numbers have a sum of 25. The average of all six numbers is 6. What is the sixth number?

a. 8
b. 10
c. 13
d. 12
e. 11

13. $52.3 \times 10^{-3} =$

a. 0.00523
b. 0.0523
c. 0.523
d. 523
e. 5.23

14. Which of the following is the result after simplifying the expression: $(7n + 3n^3 + 3) + (8n + 5n^3 + 2n^4)$?

a. $9n^4 + 15n - 2$
b. $2n^4 + 5n^3 + 15n - 2$
c. $9n^4 + 8n^3 + 15n$
d. $2n^4 + 8n^3 + 15n + 3$
e. $2n^4 + 5n^3 + 15n - 3$

15. What is the product of the following expression?

$$(4x - 8)(5x^2 + x + 6)$$

a. $20x^3 - 36x^2 + 16x - 48$
b. $6x^3 - 41x^2 + 12x + 15$
c. $9x^3 - 4x^2 - 37x - 12$
d. $2x^3 - 11x^2 - 32x + 20$
e. $20x^3 - 40x^2 + 24x - 48$

16. What is the next number in the following series: 1, 3, 6, 10, 15, 21, ...?

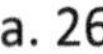

a. 26
b. 27
c. 28
d. 29
e. 30

17. How will the number 847.89632 be written if rounded to the nearest hundredth?

a. 847.90
b. 900
c. 847.89
d. 847.896
e. 847.9

18. Mom's car drove 72 miles in 90 minutes. How fast did she drive in feet per second?

a. 0.8 feet per second
b. 48.9 feet per second
c. 0.009 feet per second
d. 70.4 feet per second
e. 21.3 feet per second

19. Solve $V = lwh$ for h.

a. $lwV = h$
b. $h = \frac{V}{lw}$
c. $h = \frac{Vl}{w}$
d. $h = \frac{Vw}{l}$
e. $h = \frac{Vl}{w}$

20. Suppose $\frac{x+2}{x} = 2$. What is *x*?

a. -2
b. -1
c. 0
d. 2
e. 4

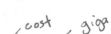

21. The phone bill is calculated each month using the equation $c = 50g + 75$. The cost of the phone bill per month is represented by c, and g represents the gigabytes of data used that month. Identify and interpret the slope of this equation.

a. 75 dollars per day
b. 75 gigabytes per day
c. 50 dollars per day
d. 50 dollars per gigabyte
e. The slope cannot be determined

22. If $\sqrt{1 + x} = 4$, what is *x*?

a. 10
b. 15
c. 20
d. 25
e. 36

23. What is the volume of a rectangular prism with the height of 3 centimeters, a width of 5 centimeters, and a depth of 11 centimeters?

a. 19 cm^3
b. 165 cm^3
c. 225 cm^3
d. 150 cm^3
e. 88 cm^3

24. What is the volume of a pyramid, with the area of the base measuring 12 $inches^2$, and the height measuring 15 inches?

a. 180 in^3
b. 90 in^3
c. 30 in^3
d. 60 in^3
e. 45 in^3

25. Which is closest to 17.8×9.9?

a. 140
b. 180
c. 200
d. 350
e. 400

26. A student gets an 85% on a test with 20 questions. How many answers did the student solve correctly?

a. 15
b. 16
c. 17
d. 18
e. 19

27. 6 is 30% of what number?

a. 18
b. 20
c. 24
d. 25
e. 26

28. Twenty is 40% of what number?

a. 60
b. 8
c. 200
d. 70
e. 50

29. What is the simplified form of the expression $1.2 \times 10^{12} \div 3.0 \times 10^{8}$?

a. 0.4×10^4
b. 4.0×10^4
c. 4.0×10^3
d. 3.6×10^{20}
e. 4.0×10^2

30. You measure the width of your door to be 36 inches. The true width of the door is 35.75 inches. What is the relative error in your measurement?

a. 0.7%
b. 0.007%
c. 0.99%
d. 0.1%
e. 7.0%

31. Using the following diagram, in terms of π, what is the total circumference of the circle?

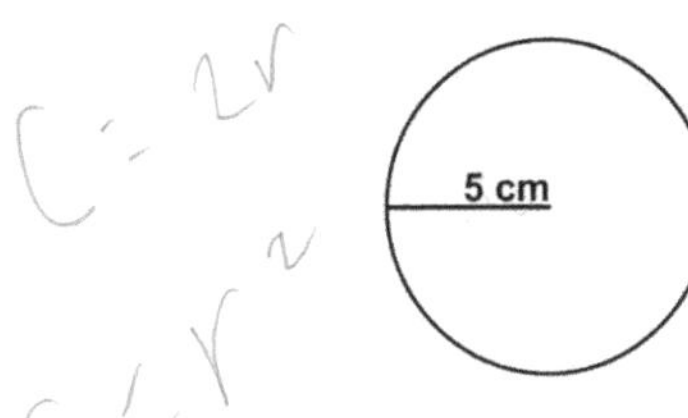

a. 2.5π cm
b. 5π cm
c. 10π cm
d. 25π cm
e. 31.4π cm

32. An angle measures 54 degrees. In order to correctly determine the measure of its complementary angle, what concept is necessary?
    a. Two complementary angles sum up to 180 degrees.
    b. Complementary angles are always acute.
    c. Two complementary angles sum up to 90 degrees.
    d. Complementary angles sum up to 360 degrees.
    e. At least one of the angles of two complementary angles must be obtuse.

33. A ball is drawn at random from a ball pit containing 8 red balls, 7 yellow balls, 6 green balls, and 5 purple balls. What's the probability that the ball drawn is yellow?
    a. $^{1}/_{26}$
    b. $^{19}/_{26}$
    c. $^{14}/_{26}$
    d. 1
    e. $^{7}/_{26}$

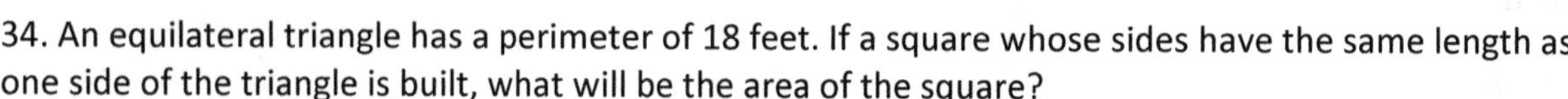

34. An equilateral triangle has a perimeter of 18 feet. If a square whose sides have the same length as one side of the triangle is built, what will be the area of the square?
    a. 6 square feet
    b. 36 square feet
    c. 256 square feet
    d. 1000 square feet
    e. 324 square feet

35. In a group of 20 men, the median weight is 180 pounds and the range is 30 pounds. If each man gains 10 pounds, which of the following would be true?
    a. The median weight will increase, and the range will remain the same.
    b. The median weight and range will both remain the same.
    c. The median weight will stay the same, and the range will increase.
    d. The median weight and range will both increase.
    e. The median weight will increase, and the range will decrease.

36. What is the slope of this line?

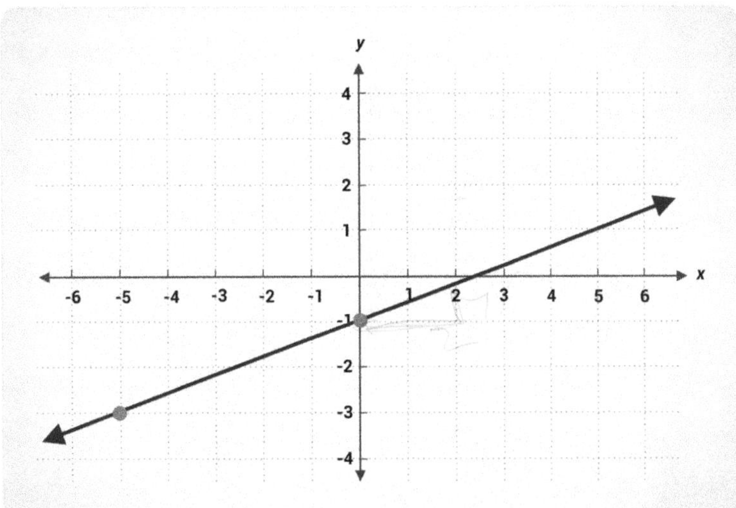

a. 2

b. $\frac{5}{2}$

c. $\frac{1}{2}$

d. $\frac{2}{5}$

e. $-\frac{5}{2}$

37. For which of the following are $x = 4$ and $x = -4$ solutions?

a. $x^2 + 16 = 0$
b. $x^2 + 4x - 4 = 0$
c. $x^2 - 2x - 2 = 0$
d. $x^2 - x - 16 = 0$
e. $x^2 - 16 = 0$

38. What are the coordinates of the point plotted on the grid?

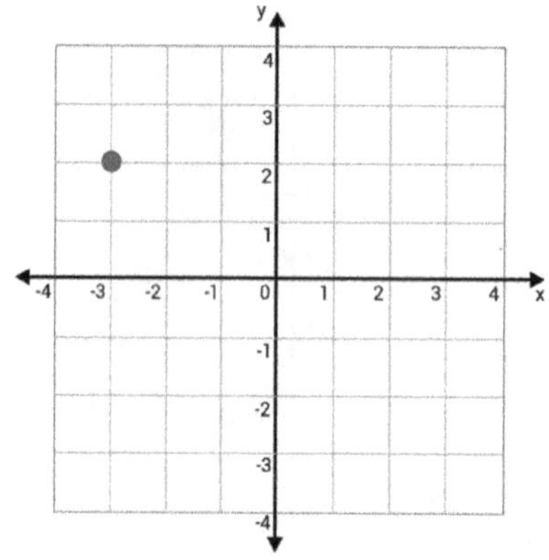

a. (-3, 2)
b. (2, -3)
c. (-3, -2)
d. (2, 3)
e. (3, 2)

39. Solve this equation:

$$9x + x - 7 = 16 + 2x$$

a. $x =$ -4

b. $x = 3$

c. $x = \frac{9}{8}$

d. $x = \frac{23}{8}$

e. $x = 4$

40. Keith's bakery had 252 customers go through its doors last week. This week, that number increased to 378. Express this increase as a percentage.

a. 26%
b. 50%
c. 35%
d. 12%
e. 42%

Answer Explanations

1. D: This problem can be solved using basic arithmetic. Xavier starts with 20 apples, then gives his sister half, so 20 divided by 2.

$$\frac{20}{2} = 10$$

He then gives his neighbor 6, so 6 is subtracted from 10.

$$10 - 6 = 4$$

Lastly, he uses ¾ of his apples to make an apple pie, so to find remaining apples, the first step is to subtract ¾ from one and then multiply the difference by 4.

$$\left(1 - \frac{3}{4}\right) \times 4 = ?$$

$$\left(\frac{4}{4} - \frac{3}{4}\right) \times 4 = ?$$

$$\left(\frac{1}{4}\right) \times 4 = 1$$

2. E: If she has used 1/3 of the paint, she has 2/3 remaining. $2\frac{1}{2}$ gallons are the same as $\frac{5}{2}$ gallons. The calculation is:

$$\frac{2}{3} \times \frac{5}{2} = \frac{5}{3} = 1\frac{2}{3} \text{ gallons}$$

3. B: $\frac{5}{2} \div \frac{1}{3} = \frac{5}{2} \times \frac{3}{1} = \frac{15}{2} = 7.5$.

4. B: The car is traveling at a speed of five meters per second. On the interval from one to three seconds, the position changes by fifteen meters. By making this change in position over time into a rate, the speed becomes ten meters in two seconds or five meters in one second.

5. A: Each bag contributes $4x + 1$ treats. The total treats will be in the form $4nx + n$ where n is the total number of bags. The total is in the form $60x + 15$, from which it is known $n = 15$.

6. E: Let a be the number of apples and b the number of bananas. The total cost is:

$$2a + 3b = 22$$

While it also known that:

$$a + b = 10$$

Using the knowledge of systems of equations, cancel the b variables by multiplying the second equation by -3. This makes the equation:

$$-3a - 3b = -30$$

Adding this to the first equation, the b values cancel to get $-a = -8$, which simplifies to $a = 8$.

7. C: To find the *y*-intercept, substitute zero for *x*, which gives us:

$$y = 0^{\frac{5}{3}} + (0-3)(0+1)$$

$$0 + (-3)(1) = -3$$

8. E: To find the average of a set of values, add the values together and then divide by the total number of values. In this case, include the unknown value of what Dwayne needs to score on his next test, in order to solve it.

$$\frac{78 + 92 + 83 + 97 + x}{5} = 90$$

Add the unknown value to the new average total, which is 5. Then multiply each side by 5 to simplify the equation, resulting in:

$$78 + 92 + 83 + 97 + x = 450$$

$$350 + x = 450$$

$$x = 100$$

Dwayne would need to get a perfect score of 100 in order to get an average of at least 90.

Test this answer by substituting back into the original formula.

$$\frac{78 + 92 + 83 + 97 + 100}{5} = 90$$

9. D: 1, 2, 3, 4, 6, 12. A given number divides evenly by each of its factors to produce an integer (no decimals). The number 5, 7, 8, 9, 10, 11 (and their opposites) do not divide evenly into 12. Therefore, these numbers are not factors.

10. A: These numbers first need to be changed to improper fractions: $\frac{11}{3} - \frac{9}{5}$. Take 15 as a common denominator:

$$\frac{11}{3} - \frac{9}{5}$$

$$\frac{55}{15} - \frac{27}{15}$$

$$\frac{28}{15}$$

$1\frac{13}{15}$ (when rewritten to get rid of the partial fraction)

11. E: Denote the width as *w* and the length as *l*. Then, $l = 3w + 5$. The perimeter is $2w + 2l = 90$. Substituting the first expression for *l* into the second equation yields:

$$2(3w + 5) + 2w = 90, \text{ or } 8w = 80, \text{ so } wl = 10$$

Putting this into the first equation, it yields:

$$l = 3(10) + 5 = 35$$

12. E: The average is calculated by adding all six numbers, then dividing by 6. The first five numbers have a sum of 25. If the total divided by 6 is equal to 6, then the total itself must be 36. The sixth number must be 36 – 25 = 11.

13. B: Multiplying by 10^{-3} means moving the decimal point three places to the left, putting in zeroes as necessary.

14. D: The expression is simplified by collecting like terms. Terms with the same variable and exponent are like terms, and their coefficients can be added, regardless of their position within the parentheses.

15. A: Finding the product means distributing one polynomial onto the other. Each term in the first must be multiplied by each term in the second. Then, like terms can be collected. Multiplying the factors yields the expression:

$$20x^3 + 4x^2 + 24x - 40x^2 - 8x - 48$$

Collecting like terms means adding the x^2 terms and adding the x terms.

The final answer after simplifying the expression is:

$$20x^3 - 36x^2 + 16x - 48$$

16. C: Each subsequent number in the sequence adds one more than the difference between the previous two. For example:

$$10 - 6 = 4, 4 + 1 = 5$$

Therefore, the next number after 10 is:

$$10 + 5 = 15$$

Going forward:

$$21 - 15 = 6, 6 + 1 = 7$$

The next number is $21 + 7 = 28$. Therefore, the difference between numbers is the set of whole numbers starting at 2: 2, 3, 4, 5, 6, 7, ….

17. A: 847.90. The hundredth-place value is located two digits to the right of the decimal point (the digit 9). The digit to the right of the place value is examined to decide whether to round up or keep the digit. In this case, the digit 6 is 5 or greater so the hundredth place is rounded up. When rounding up, if the digit to be increased is a 9, the digit to its left is increased by one and the digit in the desired place value is made a zero. Therefore, the number is rounded to 847.90.

18. D: This problem can be solved by using unit conversions. The initial units are miles per minute. The final units need to be feet per second. Converting miles to feet uses the equivalence statement 1 mile = 5,280 feet. Converting minutes to seconds uses the equivalence statement 1 minute = 60 seconds. Setting up the ratios to convert the units is shown in the following equation:

$$\frac{72\ miles}{90\ minutes} * \frac{1\ minute}{60\ seconds} * \frac{5280\ feet}{1\ mile} = 70.4\ feet\ per\ second$$

The initial units cancel out, and the new units are left.

19. B: The formula can be manipulated by dividing both the length, *l*, and the width, *w*, on both sides. The length and width will cancel on the right, leaving height by itself.

20. D: Multiply both sides by *x* to get $x + 2 = 2x$, which simplifies to $-x = -2$, or *x* = 2.

21. D: The slope from this equation is 50, and it is interpreted as the cost per gigabyte used. Since the g-value represents number of gigabytes and the equation is set equal to the cost in dollars, the slope relates these two values. For every gigabyte used on the phone, the bill goes up 50 dollars.

22. B: Start by squaring both sides to get $1 + x = 16$. Then subtract 1 from both sides to get $x = 15$.

23. B: The volume of a rectangular prism is the $length \times width \times height$, and $3cm \times 5cm \times 11cm$ is 165 cm^3. Choice *A* is not the correct answer because that is $3cm + 5cm + 11cm$. Choice *C* is not the correct answer because that is 15^2. Choice *D* is not the correct answer because that is $3cm \times 5cm \times 10cm$. Lastly, Choice *E* is incorrect because it is $(3cm + 5cm) \times 11cm$.

24. D: The volume of a pyramid is $(length \times width \times height)$, divided by 3, and (12×15), divided by 3 is 60 in^3. Choice *A* is not the correct answer because that is 12×15. Choice *B* is not the correct answer because that is (12×15), divided by 2. Choice *C* is not the correct answer because that is 15×2. Choice *E* is not the correct answer because that is (12×15), divided by 4.

25. B: Instead of multiplying these out, the product can be estimated by using $18 \times 10 = 180$. The error here should be lower than 15, since it is rounded to the nearest integer, and the numbers add to something less than 30.

26. C: 85% of a number means multiplying that number by 0.85. So, $0.85 \times 20 = \frac{85}{100} \times \frac{20}{1}$, which can be simplified to:

$$\frac{17}{20} \times \frac{20}{1} = 17$$

27. B: 30% is 3/10. The number itself must be 10/3 of 6, or:

$$\frac{10}{3} \times 6 = 10 \times 2 = 20$$

28. E: Setting up a proportion is the easiest way to represent this situation. The proportion becomes $\frac{20}{x} = \frac{40}{100}$, where cross-multiplication can be used to solve for x. The answer can also be found by observing the two fractions as equivalent, knowing that twenty is half of forty, and fifty is half of one-hundred.

29. C: Division with scientific notation can be solved by grouping the first terms together and grouping the tens together. The first terms can be divided, and the tens terms can be simplified using the rules for exponents. The initial expression becomes 0.4×10^4. This is not in scientific notation because the first number is not between 1 and 10. Shifting the decimal and subtracting one from the exponent yields 4.0×10^3.

30. A: The relative error can be found by finding the absolute error and making it a percent of the true value. The absolute error is:

$$36 - 35.75 = 0.25$$

This error is then divided by 36—the true value—to find 0.7%.

31. C: To calculate the circumference of a circle, use the formula $2\pi r$, where r equals the radius or half of the diameter of the circle. Because the radius is 5cm, the circumference is 10π cm.

32. C: The measure of two complementary angles sums up to 90 degrees. $90 - 54 = 36$. Therefore, the complementary angle is 36 degrees.

33. E: The sample space is made up of $8 + 7 + 6 + 5 = 26$ balls. The probability of pulling each individual ball is $^1/_{26}$. Since there are 7 yellow balls, the probability of pulling a yellow ball is $^7/_{26}$.

34. B: An equilateral triangle has three sides of equal length, so if the total perimeter is 18 feet, each side must be 6 feet long. A square with sides of 6 feet will have an area of $6^2 = 36$ square feet.

35. A: If each man gains 10 pounds, every original data point will increase by 10 pounds. Therefore, the man with the original median will still have the median value, but that value will be increased by 10. The smallest value and largest value will also increase by 10 and, therefore, the difference between the two won't change. The range does not change in value.

36. D: The slope is given by the change in *y* divided by the change in *x*. Specifically, it's:

$$slope = \frac{y_2 - y_1}{x_2 - x_1}$$

The first point is (-5, -3), and the second point is (0, -1). Work from left to right when identifying coordinates. Thus, the point on the left is point 1 (-5,-3) and the point on the right is point 2 (0,-1).

Now we need to just plug those numbers into the equation:

$$slope = \frac{-1 - (-3)}{0 - (-5)}$$

It can be simplified to:

$$slope = \frac{-1 + 3}{0 + 5}$$

$$slope = \frac{2}{5}$$

37. E: There are two ways to approach this problem. Each value can be substituted into each equation. Choice *A* can be eliminated, since $4^2 + 16 = 32$. Choice *B* can be eliminated, since:

$$4^2 + 4 \times 4 - 4 = 28$$

Choice *C* can be eliminated, since:

$$4^2 - 2 \cdot 4 - 2 = 6$$

But, plugging in either value into $x^2 - 16$ gives:

$$(\pm 4)^2 - 16 = \pm 16 - 16 = 0 \text{ or -32}$$

but since none of the answer choices equal -32, Choice *E* is the correct answer.

38. A: (-3, 2). The coordinates of a point are written as an ordered pair (*x*, *y*). To determine the *x*-coordinate, a line is traced directly above or below the point until reaching the *x*-axis. This step notes the value on the *x*-axis. In this case, the *x*-coordinate is -3. To determine the *y*-coordinate, a line is traced directly to the right or left of the point until reaching the y-axis, which notes the value on the *y*-axis. In this case, the *y*-coordinate is 2. Therefore, the ordered pair is written (-3, 2).

39. D:

$9x + x - 7 = 16 + 2x$	Combine $9x$ and x.
$10x - 7 = 16 + 2x$	
$10x - 7 + 7 = 16 + 2x + 7$	Add 7 to both sides to remove (-7).
$10x = 23 + 2x$	
$10x - 2x = 23 + 2x - 2x$	Subtract 2x from both sides to move it to the other side of the equation.
$8x = 23$	
$\frac{8x}{8} = \frac{23}{8}$	Divide by 8 to get x by itself.
$x = \frac{23}{8}$	

40. B: First, calculate the difference between the larger value and the smaller value.

$$378 - 252 = 126$$

To calculate this difference as a percentage of the original value, and thus calculate the percentage *increase*, divide 126 by 252, then multiply by 100 to reach the percentage 50%, answer B.

Reading

Main Ideas and Supporting Details

Topics and main ideas are critical parts of writing. The **topic** is the subject matter of the piece. An example of a topic would be *the use of cell phones in a classroom.*

The **main idea** is what the writer wants to say about that topic. A writer may make the point that the use of cell phones in a classroom is a serious problem that must be addressed in order for students to learn better. Therefore, the topic is cell phone usage in a classroom, and the main idea is that it's *a serious problem needing to be addressed*. The topic can be expressed in a word or two, but the main idea should be a complete thought.

An author will likely identify the topic immediately within the title or the first sentence of the passage. The main idea is usually presented in the introduction. In a single passage, the main idea may be identified in the first or last sentence, but it will most likely be directly stated and easily recognized by the reader. Because it is not always stated immediately in a passage, it's important that readers carefully read the entire passage to identify the main idea.

The main idea should not be confused with the thesis statement. A **thesis statement** is a clear statement of the writer's specific stance and can often be found in the introduction of a nonfiction piece. The thesis is a specific sentence (or two) that offers the direction and focus of the discussion.

In order to illustrate the main idea, a writer will use **supporting details**, which provide evidence or examples to help make a point. Supporting details are typically found in nonfiction pieces that seek to inform or persuade the reader.

In the example of cell phone usage in the classroom, where the author's main idea is to show the seriousness of this problem and the need to "unplug", supporting details would be critical for effectively making that point. Supporting details used here might include statistics on a decline in student focus and studies showing the impact of digital technology usage on students' attention spans. The author could also include testimonies from teachers surveyed on the topic.

It's important that readers evaluate the author's supporting details to be sure that they are credible, provide evidence of the author's point, and directly support the main idea. Although shocking statistics grab readers' attention, their use may provide ineffective information in the piece. Details like this are crucial to understanding the passage and evaluating how well the author presents his or her argument and evidence.

Also remember that when most authors write, they want to make a point or send a message. This point or message of a text is known as the theme. Authors may state themes explicitly, like in *Aesop's Fables*. More often, especially in modern literature, readers must infer the theme based on text details. Usually after carefully reading and analyzing an entire text, the theme emerges. Typically, the longer the piece, the more themes you will encounter, though often one theme dominates the rest, as evidenced by the author's purposeful revisiting of it throughout the passage.

Recognizing the Structure of Texts in Various Formats

Writing can be classified under four passage types: narrative, expository, descriptive (sometimes called technical), and persuasive. Though these types are not mutually exclusive, one form tends to dominate the rest. By recognizing the *type* of passage you're reading, you gain insight into *how* you should read. If you're reading a narrative, you can assume the author intends to entertain, which means you may skim the text without losing meaning. A technical document might require a close read, because skimming the passage might cause the reader to miss salient details.

1. **Narrative** writing, at its core, is the art of storytelling. For a narrative to exist, certain elements must be present. First, it must have characters While many characters are human, characters could be defined as anything that thinks, acts, and talks like a human. For example, many recent movies, such as *Lord of the Rings* and *The Chronicles of Narnia*, include animals, fantastical creatures, and even trees that behave like humans. Second, it must have a plot or sequence of events. Typically, those events follow a standard plot diagram, but recent trends start *in medias res* or in the middle (near the climax). In this instance, foreshadowing and flashbacks often fill in plot details. Finally, along with characters and a plot, there must also be conflict. Conflict is usually divided into two types: internal and external. Internal conflict indicates the character is in turmoil and is presented through the character's thoughts. External conflicts are visible. Types of external conflict include a person versus nature, another person, or society.

2. **Expository** writing is detached and to the point. Since expository writing is designed to instruct or inform, it usually involves directions and steps written in second person ("you" voice) and lacks any persuasive or narrative elements. Sequence words such as *first, second,* and *third,* or *in the first place, secondly,* and *lastly* are often given to add fluency and cohesion. Common examples of expository writing include instructor's lessons, cookbook recipes, and repair manuals.

3. Due to its empirical nature, **technical** writing is filled with steps, charts, graphs, data, and statistics. The goal of technical writing is to advance understanding in a field through the scientific method. Experts such as teachers, doctors, or mechanics use words unique to the profession in which they operate. These words, which often incorporate acronyms, are called **jargon**. Technical writing is a type of expository writing but is not meant to be understood by the general public. Instead, technical writers assume readers have received a formal education in a particular field of study and need no explanation as to what the jargon means. Imagine a doctor trying to understand a diagnostic reading for a car or a mechanic trying to interpret lab results. Only professionals with proper training will fully comprehend the text.

4. **Persuasive** writing is designed to change opinions and attitudes. The topic, stance, and arguments are found in the thesis, positioned near the end of the introduction. Later supporting paragraphs offer relevant quotations, paraphrases, and summaries from primary or secondary sources, which are then interpreted, analyzed, and evaluated. The goal of persuasive writers is not to stack quotes, but to develop original ideas by using sources as a starting point. Good persuasive writing makes powerful arguments with valid sources and thoughtful analysis. Poor persuasive writing is riddled with bias and logical fallacies. Sometimes logical and illogical arguments are sandwiched together in the same piece. Therefore, readers should display skepticism when reading persuasive arguments.

Inferences

Simply put, an **inference** is an educated guess drawn from evidence, logic, and reasoning. The key to making inferences is identifying clues within a passage, and then using common sense to arrive at a reasonable conclusion. It can be considered "reading between the lines."

One way to make an inference is to look for main topics. When doing so, readers should pay particular attention to any titles, headlines, or opening statements made by the author. Topic sentences or repetitive ideas can be clues in gleaning inferred ideas. For example, if a passage contains the phrase *DNA testing, while some consider it infallible, is an inherently flawed technique,* the test taker can infer the rest of the passage will contain information that points to DNA testing's infallibility.

The test taker may be asked to make an inference based on prior knowledge, but may also be asked to make predictions based on new ideas. For example, the test taker may have no prior knowledge of DNA other than its genetic property to replicate. However, if the reader is given passages on the flaws of DNA testing with enough factual evidence, the test taker may arrive at the inferred conclusion that the author does not support the infallibility of DNA testing in all identification cases.

When making inferences, it is important to remember that the critical thinking process involved must be fluid and open to change. While readers may infer an idea from a main topic, general statement, or other clues, they must be open to receiving new information within a particular passage. New ideas presented by an author may require the test taker to alter an inference. Similarly, when asked questions that require making an inference, it's important for readers to read the entire test passage and all of the answer options. Often, a test taker will need to refine a general inference based on new ideas that may be presented within the test itself.

Context of a Word or Phrase

A **context clue** is a hint that an author provides to the reader in order to help define difficult or unique words. When reading a passage, test takers should take note of any unfamiliar words, and then examine the sentence around those words to look for clues to the word meanings.

Let's look at an example:

> He faced a *conundrum* in making this decision. He felt as if he had come to a crossroads. This was truly a puzzle, and what he did next would determine the course of his future.

The word *conundrum* may be unfamiliar to the reader. However, by looking at context clues, the reader should be able to determine its meaning. In this passage, context clues include the idea of making a decision and of being unsure. Furthermore, the author restates the definition of conundrum in using the word *puzzle* as a synonym. Therefore, the reader should be able to determine that the definition of the word *conundrum* is a difficult puzzle.

Similarly, a reader can determine difficult vocabulary by identifying antonyms. Let's look at an example:

> Her *gregarious* nature was completely opposite of her twin's, who was shy, retiring, and socially nervous.

The word *gregarious* may be unfamiliar. However, by looking at the surrounding context clues, the reader can determine that *gregarious* does not mean shy. The twins' personalities are being contrasted. Therefore, *gregarious* must mean sociable, or something similar to it.

At times, an author will provide contextual clues through a cause and effect relationship. Look at the next sentence as an example:

> The athletes were *elated* when they won the tournament; unfortunately, their off-court antics caused them to forfeit the win.

The word *elated* may be unfamiliar to the reader. However, the author defines the word by presenting a cause and effect relationship. The athletes were so elated at the win that their behavior went overboard and they had to forfeit. In this instance, *elated* must mean something akin to overjoyed, happy, and overexcited.

Cause and effect is one technique authors use to demonstrate relationships. A **cause** is why something happens. The **effect** is what happens as a result. For example, a reader may encounter text such as *Because he was unable to sleep, he was often restless and irritable during the day.* The cause is insomnia due to lack of sleep. The effect is being restless and irritable. When reading for a cause and effect relationship, look for words such as "if", "then", "such", and "because." By using cause and effect, an author can describe direct relationships, and convey an overall theme, particularly when taking a stance on their topic.

An author can also provide contextual clues through comparison and contrast. Let's look at an example:

> Her torpid state caused her parents, and her physician, to worry about her seemingly sluggish well-being.

The word *torpid* is probably unfamiliar to the reader. However, the author has compared *torpid* to a state of being and, moreover, one that's worrisome. Therefore, the reader should be able to determine that *torpid* is not a positive, healthy state of being. In fact, through the use of comparison, it means sluggish. Similarly, an author may contrast an unfamiliar word with an idea. In the sentence *Her torpid state was completely opposite of her usual, bubbly self,* the meaning of *torpid*, or sluggish, is contrasted with the words *bubbly self*.

A reader should be able to critically assess and determine unfamiliar word meanings through the use of an author's context clues in order to fully comprehend difficult text passages.

Author's Purpose

No matter the genre or format, all authors are writing to persuade, inform, entertain, or express feelings. Often, these purposes are blended, with one dominating the rest. It's useful to learn to recognize the author's intent.

Persuasive writing is used to persuade or convince readers of something. It often contains two elements: the argument and the counterargument. The argument takes a stance on an issue, while the counterargument pokes holes in the opposition's stance. Authors rely on logic, emotion, and writer credibility to persuade readers to agree with them. If readers are opposed to the stance before reading, they are unlikely to adopt that stance. However, those who are undecided or committed to the same stance are more likely to agree with the author.

Informative writing tries to teach or inform. Workplace manuals, instructor lessons, statistical reports and cookbooks are examples of informative texts. Informative writing is usually based on facts and is often void of emotion and persuasion. Informative texts generally contain statistics, charts, and graphs. Although most informative texts lack a persuasive agenda, readers must examine the text carefully to determine whether one exists within a given passage.

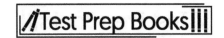

Stories or narratives are designed to entertain. When you go to the movies, you often want to escape for a few hours, not necessarily to think critically. Entertaining writing is designed to delight and engage the reader. However, sometimes this type of writing can be woven into more serious materials, such as persuasive or informative writing to hook the reader before transitioning into a more scholarly discussion.

Emotional writing works to evoke the reader's feelings, such as anger, euphoria, or sadness. The connection between reader and author is an attempt to cause the reader to share the author's intended emotion or tone. Sometimes in order to make a piece more poignant, the author simply wants readers to feel the same emotions that the author has felt. Other times, the author attempts to persuade or manipulate the reader into adopting his stance. While it's okay to sympathize with the author, be aware of the individual's underlying intent.

Author's Attitude and Tone

An author's **tone** is the use of particular words, phrases, and writing style to convey an overall meaning. Tone expresses the author's attitude towards a particular topic. For example, a historical reading passage may begin like the following:

> The presidential election of 1960 ushered in a new era, a new Camelot, a new phase of forward thinking in U.S. politics that embraced brash action, unrest, and responded with admirable leadership.

From this opening statement, a reader can draw some conclusions about the author's attitude towards President John F. Kennedy. Furthermore, the reader can make additional, educated guesses about the state of the Union during the 1960 presidential election. By close reading, the test taker can determine that the repeated use of the word *new* and words such as *admirable leadership* indicate the author's tone of admiration regarding the President's boldness. In addition, the author assesses that the era during President Kennedy's administration was problematic through the use of the words *brash action* and *unrest.* Therefore, if a test taker encountered a test question asking about the author's use of tone and their assessment of the Kennedy administration, the test taker should be able to identify an answer indicating admiration. Similarly, if asked about the state of the Union during the 1960s, a test taker should be able to correctly identify an answer indicating political unrest.

Generally, parts of speech that indicate attitude will also indicate tone. When identifying an author's tone, the following list of words, while not exhaustive, may be helpful:

- Comical
- Angry
- Ambivalent
- Scary
- Lyrical
- Matter-of-fact

- Judgmental
- Sarcastic
- Malicious
- Objective
- Pessimistic
- Patronizing
- Gloomy
- Instructional
- Satirical
- Formal
- Casual

Critical Thinking Skills

It's important to read any piece of writing critically. The goal is for the reader to discover the point and purpose of what the author is writing about through analysis. It's also crucial for the reader to establish the point or stance the author has taken on the topic of the piece. After determining the author's perspective, readers can then more effectively develop their own viewpoints on the subject of the piece.

It is important to distinguish between fact and opinion when reading a piece of writing. A **fact** is information that can be proven true. If information can be disproved, it is not a fact. For example, water freezes at or below thirty-two degrees Fahrenheit. An argument stating that water freezes at seventy degrees Fahrenheit cannot be supported by data and is therefore not a fact. Facts tend to be associated with science, mathematics, and statistics. **Opinions** are information open to debate. Opinions are often tied to subjective concepts like equality, morals, and rights. They can also be controversial.

Authors often use words like *think, feel, believe,* or *in my opinion* when expressing opinion, but these words won't always appear in an opinion piece, especially if it is formally written. An author's opinion may be backed up by facts, which gives it more credibility, but that opinion should not be taken as fact. A critical reader should be suspect of an author's opinion, especially if it is only supported by other opinions.

Fact	Opinion
There are 9 innings in a game of baseball.	Baseball games run too long.
James Garfield was assassinated on July 2, 1881.	James Garfield was a good president.
McDonalds has stores in 120 countries.	McDonalds has the best hamburgers.

Evaluating an Argument and its Specific Claims

It's important to evaluate the author's supporting details to be sure that the details are credible, provide evidence of the author's point, and directly support the main idea. Although shocking statistics grab readers' attention, their use could be ineffective in the piece. Details like this are crucial to understanding the passage and evaluating how well the author presents their argument and evidence.

Readers draw **conclusions** about what an author has presented. This helps them better understand what the writer has intended to communicate and whether or not they agree with what the author has offered. There are a few ways to determine a logical conclusion, but careful reading is the most important. It's helpful to read a passage a few times, noting details that seem important to the piece.

Sometimes, readers arrive at a conclusion that is different than what the writer intended or come up with more than one conclusion.

Evidence

When an author presents **facts**, such as statistics or data, readers should be able to check those facts and make sure they are accurate. When authors use **opinion**, they are sharing their own thoughts and feelings about a subject.

Textual evidence within the details helps readers draw a conclusion about a passage. **Textual evidence** refers to information—facts and examples that support the main point. Textual evidence will likely come from outside sources and can be in the form of quoted or paraphrased material. In order to draw a conclusion from evidence, it's important to examine the credibility and validity of that evidence as well as how (and if) it relates to the main idea.

Credibility

Critical readers examine the facts used to support an author's argument. They check the facts against other sources to be sure those facts are correct. They also check the validity of the sources used to be sure those sources are credible, academic, and/or peer- reviewed. Consider that when an author uses another person's opinion to support their argument, even if it is an expert's opinion, it is still only an opinion and should not be taken as fact. A strong argument uses valid, measurable facts to support ideas. Even then, the reader may disagree with the argument as it may be rooted in their personal beliefs.

An authoritative argument may use the facts to sway the reader. In the example of global warming, many experts differ in their opinions of what alternative fuels can be used to aid in offsetting it. Because of this, a writer may choose to only use the information and expert opinion that supports their viewpoint.

Appeal to Emotion

An author's argument might also appeal to readers' emotions, perhaps by including personal stories and **anecdotes** (a short narrative of a specific event).

The next example presents an appeal to emotion. By sharing the personal anecdote of one student and speaking about emotional topics like family relationships, the author invokes the reader's empathy in asking them to reconsider the school rule.

Our school should abolish its current ban on cell phone use on campus. If they aren't able to use their phones during the school day, many students feel isolated from their loved ones. For example, last semester, one student's grandmother had a heart attack in the morning. However, because he couldn't use his cell phone, the student didn't know about his grandmother's accident until the end of the day—when she had already passed away and it was too late to say goodbye. By preventing students from contacting their friends and family, our school is placing undue stress and anxiety on students.

Counterarguments

If an author presents a differing opinion or a **counterargument** in order to refute it, the reader should consider how and why this information is being presented. It is meant to strengthen the original argument and shouldn't be confused with the author's intended conclusion, but it should also be considered in the reader's final evaluation. On the contrary, sometimes authors will concede to an opposing argument by recognizing the validity the other side has to offer. A concession will allow readers to see both sides of the argument in an unbiased light, thereby increasing the credibility of the author.

Authors can also reflect **bias** if they ignore an opposing viewpoint or present their side in an unbalanced way. A strong argument considers the opposition and finds a way to refute it. Critical readers should look for an unfair or one-sided presentation of the argument and be skeptical, as a bias may be present. Even if this bias is unintentional, if it exists in the writing, the reader should be wary of the validity of the argument.

Making Predictions

There are a few ways for readers to engage actively with the text, such as making inferences and predictions. An **inference** refers to a point that is implied (as opposed to directly-stated) by the evidence presented:

> Bradley packed up all of the items from his desk in a box and said goodbye to his coworkers for the last time.

Although it is not directly stated, from this sentence, readers can infer that Bradley is leaving his job. It's necessary to use inference in order to draw conclusions about the meaning of a passage. Authors make implications through character dialogue, thoughts, effects on others, actions, and looks. Like in life, readers must assemble all the clues to form a complete picture.

When making an inference about a passage, it's important to rely only on the information that is provided in the text itself. This helps readers ensure that their conclusions are valid.

Readers will also find themselves making predictions when reading a passage or paragraph. **Predictions** are guesses about what's going to happen next. Readers can use prior knowledge to help make accurate predictions. Prior knowledge is best utilized when readers make links between the current text, previously read texts, and life experiences. Some texts use suspense and foreshadowing to captivate readers:

> A cat darted across the street just as the car came careening around the curve.

One unfortunate prediction might be that the car will hit the cat. Of course, predictions aren't always accurate, so it's important to read carefully to the end of the text to determine the accuracy of predictions.

Comparing and Contrasting Themes from Print and Other Sources

Identifying Theme or Central Message

The **theme** is the central message of a fictional work, whether that work is structured as prose, drama, or poetry. It is the heart of what an author is trying to say to readers through the writing, and theme is largely conveyed through literary elements and techniques.

In literature, a theme can often be determined by considering the over-arching narrative conflict within the work. Though there are several types of conflicts and several potential themes within them, the following are the most common:

- Individual against the self—relevant to themes of self-awareness, internal struggles, pride, coming of age, facing reality, fate, free will, vanity, loss of innocence, loneliness, isolation, fulfillment, failure, and disillusionment
- Individual against nature— relevant to themes of knowledge vs. ignorance, nature as beauty, quest for discovery, self-preservation, chaos and order, circle of life, death, and destruction of beauty
- Individual against society— relevant to themes of power, beauty, good, evil, war, class struggle, totalitarianism, role of men/women, wealth, corruption, change vs. tradition, capitalism, destruction, heroism, injustice, and racism
- Individual against another individual— relevant to themes of hope, loss of love or hope, sacrifice, power, revenge, betrayal, and honor

For example, in Hawthorne's *The Scarlet Letter*, one possible narrative conflict could be the individual against the self, with a relevant theme of internal struggles. This theme is alluded to through characterization—Dimmesdale's moral struggle with his love for Hester and Hester's internal struggles with the truth and her daughter, Pearl. It's also alluded to through plot—Dimmesdale's suicide and Hester helping the very townspeople who initially condemned her.

Sometimes, a text can convey a **message** or **universal lesson**—a truth or insight that the reader infers from the text, based on analysis of the literary and/or poetic elements. This message is often presented as a statement. For example, a potential message in Shakespeare's *Hamlet* could be "Revenge is what ultimately drives the human soul." This message can be immediately determined through plot and characterization in numerous ways, but it can also be determined through the setting of Norway, which is bordering on war.

How Authors Develop Theme

Authors employ a variety of techniques to present a theme. They may compare or contrast characters, events, places, ideas, or historical or invented settings to speak thematically. They may use analogies, metaphors, similes, allusions, or other literary devices to convey the theme. An author's use of diction, syntax, and tone can also help convey the theme. Authors will often develop themes through the development of characters, use of the setting, repetition of ideas, use of symbols, and through contrasting value systems. Authors of both fiction and nonfiction genres will use a variety of these techniques to develop one or more themes.

Regardless of the literary genre, there are commonalities in how authors, playwrights, and poets develop themes or central ideas.

Authors often do research, the results of which contributes to theme. In prose fiction and drama, this research may include real historical information about the setting the author has chosen or include elements that make fictional characters, settings, and plots seem realistic to the reader. In nonfiction, research is critical since the information contained within this literature must be accurate and, moreover, accurately represented.

In fiction, authors present a narrative conflict that will contribute to the overall theme. This conflict, in fiction texts, may involve the storyline itself and some trouble within characters that needs resolution. In nonfiction, this conflict may be an explanation or commentary on factual people and events.

Authors will sometimes use character motivation to convey theme, such as in the example from *Hamlet* regarding revenge. In fiction, the characters an author creates will think, speak, and act in ways that effectively convey the theme to readers. In nonfiction, the characters are factual, as in a biography, but authors pay particular attention to presenting those motivations to make them clear to readers.

Authors also use literary devices as a means of conveying theme. For example, the use of moon symbolism in Mary Shelley's *Frankenstein* is significant as its phases can be compared to the phases that the Creature undergoes as he struggles with his identity.

The selected point of view can also contribute to a work's theme. The use of first-person point of view in a fiction or nonfiction work engages the reader's response differently than third person point of view. The central idea or theme from a first-person narrative may differ from a third-person limited text.

In literary nonfiction, authors usually identify the purpose of their writing, which differs from fiction, where the general purpose is to entertain. The purpose of nonfiction is usually to inform, persuade, or entertain the audience. The stated purpose of a non-fiction text will drive how the central message or theme, if applicable, is presented.

Authors identify an audience for their writing, which is critical in shaping the theme of the work. For example, the audience for J.K. Rowling's *Harry Potter* series would be different than the audience for a biography of George Washington. The audience an author chooses to address is closely tied to the purpose of the work. The choice of an audience also drives the choice of language and level of diction an author uses. Ultimately, the intended audience determines the level to which that subject matter is presented and the complexity of the theme.

Cultural Influence on Themes

Regardless of culture, place, or time, certain themes are universal to the human condition. Because all humans experience certain feelings and engage in similar experiences—birth, death, marriage, friendship, finding meaning, etc.—certain themes span cultures. However, different cultures have different norms and general beliefs concerning these themes. For example, the theme of maturing and crossing from childhood to adulthood is a global theme; however, the literature from one culture might imply that this happens in someone's twenties, while another culture's literature might imply that it happens in the early teenage years.

It's important for the reader to be aware of these differences. Readers must avoid being **ethnocentric,** which means believing the aspects of one's own culture to be superior to that of other cultures.

Identifying Rhetorical Devices

If one feels strongly about a subject, or has a passion for it, they choose strong words and phrases. Think of the types of rhetoric (or language) our politicians use. Each word, phrase, and idea is carefully crafted to elicit a response. Hopefully, that response is one of agreement to a certain point of view, especially among voters. Authors use the same types of language to achieve the same results. For example, the word "bad" has a certain connotation, but the words "horrid," "repugnant," and "abhorrent" paint a far better picture for the reader. They are more precise. They're interesting to read and they should all illicit stronger feelings in the reader than the word "bad." An author generally uses other devices beyond mere word choice to persuade, convince, entertain, or otherwise engage a reader.

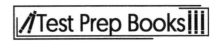

Rhetorical devices are those elements an author utilizes in painting sensory, and hopefully persuasive ideas to which a reader can relate. They are numerable. Test takers will likely encounter one or more standardized test questions addressing varying rhetorical devices. This study guide will address the more common types: alliteration, irony, metaphor, simile, hyperbole, allegory, imagery, onomatopoeia, and personification, providing examples of each.

Alliteration is a device that uses repetitive beginning sounds in words to appeal to the reader. Classic tongue twisters are a great example of alliteration. *She sells sea shells down by the sea shore* is an extreme example of alliteration. Authors will use alliterative devices to capture a reader's attention. It's interesting to note that marketing also utilizes alliteration in the same way. A reader will likely remember products that have the brand name and item starting with the same letter. Similarly, many songs, poems, and catchy phrases use this device. It's memorable. Use of alliteration draws a reader's attention to ideas that an author wants to highlight.

Irony is a device that authors use when pitting two contrasting items or ideas against each other in order to create an effect. It's frequently used when an author wants to employ humor or convey a sarcastic tone. Additionally, it's often used in fictional works to build tension between characters, or between a particular character and the reader. An author may use **verbal irony** (sarcasm), **situational irony** (where actions or events have the opposite effect than what's expected), and **dramatic irony** (where the reader knows something a character does not). Examples of irony include:

- Dramatic Irony: An author describing the presence of a hidden killer in a murder mystery, unbeknownst to the characters but known to the reader.
- Situational Irony: An author relating the tale of a fire captain who loses her home in a five-alarm conflagration.
- Verbal Irony: This is where an author or character says one thing but means another. For example, telling a police officer "Thanks a lot" after receiving a ticket.

Metaphor is a device that uses a figure of speech to paint a visual picture of something that is not literally applicable. Authors relate strong images to readers, and evoke similar strong feelings using metaphors. Most often, authors will mention one thing in comparison to another more familiar to the reader. It's important to note that metaphors do not use the comparative words "like" or "as." At times, metaphors encompass common phrases such as clichés. At other times, authors may use mixed metaphors in making identification between two dissimilar things.

Examples of metaphors include:

- An author describing a character's anger as *a flaming sheet of fire.*
- An author relating a politician as having been a folding chair under close questioning.
- A novel's character telling another character to *take a flying hike.*
- Shakespeare's assertion that *all the world's a stage.*

Simile is a device that compares two dissimilar things using the words "like" and "as." When using similes, an author tries to catch a reader's attention and use comparison of unlike items to make a point. Similes are commonly used and often develop into figures of speech and catch phrases.

Examples of similes include:

- An author describing a character as having a complexion like a faded lily.
- An investigative journalist describing his interview subject as being like cold steel and with a demeanor hard as ice.
- An author asserting the current political arena is just like a three-ring circus and as dry as day old bread.

Similes and metaphors can be confusing. When utilizing simile, an author will state one thing is like another. A metaphor states one thing is another. An example of the difference would be if an author states a character is *just like a fierce tiger and twice as angry,* as opposed to stating the character *is a fierce tiger and twice as angry.*

Hyperbole is simply an exaggeration that is not taken literally. A potential test taker will have heard or employed hyperbole in daily speech, as it is a common device we all use. Authors will use hyperbole to draw a reader's eye toward important points and to illicit strong emotional and relatable responses.

Examples of hyperbole include:

- An author describing a character as being as big as a house and twice the circumference of a city block.
- An author stating the city's water problem as being old as the hills and more expensive than a king's ransom in spent tax dollars.
- A journalist stating the mayoral candidate died of embarrassment when her tax records were made public.

Allegories are stories or poems with hidden meanings, usually a political or moral one. Authors will frequently use allegory when leading the reader to a conclusion. Allegories are similar to parables, symbols, and analogies. Often, an author will employ the use of allegory to make political, historical, moral, or social observations. As an example, Jonathan Swift's work *Gulliver's Travels into Several Remote Nations of the World* is an allegory in and of itself. The work is a political allegory of England during Jonathan Swift's lifetime. Set in the travel journal style plot of a giant amongst smaller people, and a smaller Gulliver amongst the larger, it is a commentary on Swift's political stance of existing issues of his age. Many fictional works are entire allegories in and of themselves. George Orwell's *Animal Farm* is a story of animals that conquer man and form their own farm society with swine at the top; however, it is not a literal story in any sense. It's Orwell's political allegory of Russian society during and after the Communist revolution of 1917. Other examples of allegory in popular culture include:

- Aesop's fable "The Tortoise and the Hare," which teaches readers that being steady is more important than being fast and impulsive.
- The popular *Hunger Games* by Suzanne Collins that teaches readers that media can numb society to what is truly real and important.
- Dr. Seuss's *Yertle the Turtle* which is a warning against totalitarianism and, at the time it was written, against the despotic rule of Adolf Hitler.

Imagery is a rhetorical device that an author employs when they use visual, or descriptive, language to evoke a reader's emotion. Use of imagery as a rhetorical device is broader in scope than this study guide addresses, but in general, the function of imagery is to create a vibrant scene in the reader's imagination and, in turn, tease the reader's ability to identify through strong emotion and sensory experience. In the simplest of terms, imagery, as a rhetoric device, beautifies literature.

An example of poetic imagery is below:

> Pain has an element of blank
>
> It cannot recollect
>
> When it began, or if there were
>
> A day when it was not.
>
> It has no future but itself,
>
> Its infinite realms contain
>
> Its past, enlightened to perceive
>
> New periods of pain.

In the above poem, Emily Dickenson uses strong imagery. Pain is equivalent to an "element of blank" or of nothingness. Pain cannot recollect a beginning or end, as if it was a person (see *personification* below). Dickenson appeals to the reader's sense of a painful experience by discussing the unlikelihood that discomfort sees a future, but does visualize a past and present. She simply indicates that pain, through the use of imagery, is cyclical and never ending. Dickenson's theme is one of painful depression and it is through the use of imagery that she conveys this to her readers.

Onomatopoeia is the author's use of words that create sound. Words like *pop* and *sizzle* are examples of onomatopoeia. When an author wants to draw a reader's attention in an auditory sense, they will use onomatopoeia. An author may also use onomatopoeia to create sounds as interjection or commentary.

Examples include:

- An author describing a cat's vocalization as a *chirp echoing throughout the empty cabin*.
- A description of a campfire as *crackling and whining against its burning green wood*.
- An author relating the sound of a car accident as *metallic screeching against crunching asphalt*.
- A description of an animal roadblock as being *a symphonic melody of groans, baas, and moans*.

Personification is a rhetorical device that an author uses to attribute human qualities to inanimate objects or animals. Once again, this device is useful when an author wants the reader to strongly relate to an idea. As in the example of George Orwell's *Animal Farm*, many of the animals are given the human abilities to speak, reason, apply logic, and otherwise interact as humans do. This helps the reader see how easily it is for any society to segregate into the haves and the have-nots through the manipulation of power. Personification is a device that enables the reader to empathize through human experience.

Examples of personification include:

- An author describing the wind as *whispering through the trees.*
- A description of a stone wall as being *a hardened, unmovable creature* made of cement and brick.
- An author attributing a city building as having *slit eyes and an unapproachable, foreboding façade.*
- An author describing spring as *a beautiful bride, blooming in white, ready for summer's matrimony.*

When identifying rhetorical devices, readers should look for words and phrases that capture one's attention and make note of the author's use of comparison between the inanimate and the animate. They should consider words that make the reader feel sounds and envision imagery, and pay attention to the rhythm of fluid sentences and to the use of words that evoke emotion. The ability to identify rhetorical devices is another step in achieving successful reading comprehension and in being able to correctly answer standardized questions related to those devices.

Practice Questions

Questions 1–6 are based upon the following passage:

This excerpt is an adaptation of Jonathan Swift's *Gulliver's Travels into Several Remote Nations of the World.*

> My gentleness and good behaviour had gained so far on the emperor and his court, and indeed upon the army and people in general, that I began to conceive hopes of getting my liberty in a short time. I took all possible methods to cultivate this favourable disposition. The natives came, by degrees, to be less apprehensive of any danger from me. I would sometimes lie down, and let five or six of them dance on my hand; and at last the boys and girls would venture to come and play at hide-and-seek in my hair. I had now made a good progress in understanding and speaking the language. The emperor had a mind one day to entertain me with several of the country shows, wherein they exceed all nations I have known, both for dexterity and magnificence. I was diverted with none so much as that of the rope-dancers, performed upon a slender white thread, extended about two feet, and twelve inches from the ground. Upon which I shall desire liberty, with the reader's patience, to enlarge a little.
>
> This diversion is only practised by those persons who are candidates for great employments, and high favour at court. They are trained in this art from their youth, and are not always of noble birth, or liberal education. When a great office is vacant, either by death or disgrace (which often happens,) five or six of those candidates petition the emperor to entertain his majesty and the court with a dance on the rope; and whoever jumps the highest, without falling, succeeds in the office. Very often the chief ministers themselves are commanded to show their skill, and to convince the emperor that they have not lost their faculty. Flimnap, the treasurer, is allowed to cut a caper on the straight rope, at least an inch higher than any other lord in the whole empire. I have seen him do the summerset several times together, upon a trencher fixed on a rope which is no thicker than a common packthread in England. My friend Reldresal, principal secretary for private affairs, is, in my opinion, if I am not partial, the second after the treasurer; the rest of the great officers are much upon a par.

1. Which of the following statements best summarize the central purpose of this text?
   a. Gulliver details his fondness for the archaic yet interesting practices of his captors.
   b. Gulliver conjectures about the intentions of the aristocratic sector of society.
   c. Gulliver becomes acquainted with the people and practices of his new surroundings.
   d. Gulliver's differences cause him to become penitent around new acquaintances.

2. What is the word *principal* referring to in the following text?

   > My friend Reldresal, principal secretary for private affairs, is, in my opinion, if I am not partial, the second after the treasurer; the rest of the great officers are much upon a par.

   a. Primary or chief
   b. An acolyte
   c. An individual who provides nurturing
   d. One in a subordinate position

3. What can the reader infer from this passage?

I would sometimes lie down, and let five or six of them dance on my hand; and at last the boys and girls would venture to come and play at hide-and-seek in my hair.

a. The children tortured Gulliver.
b. Gulliver traveled because he wanted to meet new people.
c. Gulliver is considerably larger than the children who are playing around him.
d. Gulliver has a genuine love and enthusiasm for people of all sizes.

4. What is the significance of the word *mind* in the following passage?

The emperor had a mind one day to entertain me with several of the country shows, wherein they exceed all nations I have known, both for dexterity and magnificence.

a. The ability to think
b. A collective vote
c. A definitive decision
d. A mythological question

5. Which of the following assertions does not support the fact that games are a commonplace event in this culture?

a. My gentlest and good behavior . . . short time.
b. They are trained in this art from their youth . . . liberal education.
c. Very often the chief ministers themselves are commanded to show their skill . . . not lost their faculty.
d. Flimnap, the treasurer, is allowed to cut a caper on the straight rope . . . higher than any other lord in the whole empire.

6. How do the roles of Flimnap and Reldresal serve as evidence of the community's emphasis regarding the correlation between physical strength and leadership abilities?

a. Only children used Gulliver's hands as a playground.
b. The two men who exhibited superior abilities held prominent positions in the community.
c. Only common townspeople, not leaders, walk the straight rope.
d. No one could jump higher than Gulliver.

Questions 7–12 are based upon the following passage:

This excerpt is adaptation of Robert Louis Stevenson's *The Strange Case of Dr. Jekyll and Mr. Hyde.*

"Did you ever come across a protégé of his—one Hyde?" He asked.

"Hyde?" repeated Lanyon. "No. Never heard of him. Since my time."

That was the amount of information that the lawyer carried back with him to the great, dark bed on which he tossed to and fro until the small hours of the morning began to grow large. It was a night of little ease to his toiling mind, toiling in mere darkness and besieged by questions.

Six o'clock struck on the bells of the church that was so conveniently near to Mr. Utterson's dwelling, and still he was digging at the problem. Hitherto it had touched him on the intellectual side alone; but; but now his imagination also was engaged, or rather enslaved; and as he lay and tossed in the gross darkness of the night in the curtained room, Mr. Enfield's tale went by before his mind in a scroll of lighted pictures. He would be aware of the great field of lamps in a nocturnal city; then of the figure of a man walking swiftly; then of a child running from the doctor's; and then these met, and that human Juggernaut trod the child down and passed on regardless of her screams. Or else he would see a room in a rich house, where his friend lay asleep, dreaming and smiling at his dreams; and then the door of that room would be opened, the curtains of the bed plucked apart, the sleeper recalled, and, lo! There would stand by his side a figure to whom power was given, and even at that dead hour he must rise and do its bidding. The figure in these two phrases haunted the lawyer all night; and if at anytime he dozed over, it was but to see it glide more stealthily through sleeping houses, or move the more swiftly, and still the more smoothly, even to dizziness, through wider labyrinths of lamplighted city, and at every street corner crush a child and leave her screaming. And still the figure had no face by which he might know it; even in his dreams it had no face, or one that baffled him and melted before his eyes; and thus there it was that there sprung up and grew apace in the lawyer's mind a singularly strong, almost an inordinate, curiosity to behold the features of the real Mr. Hyde. If he could but once set eyes on him, he thought the mystery would lighten and perhaps roll altogether away, as was the habit of mysterious things when well examined. He might see a reason for his friend's strange preference or bondage, and even for the startling clauses of the will. And at least it would be a face worth seeing: the face of a man who was without bowels of mercy: a face which had but to show itself to raise up, in the mind of the unimpressionable Enfield, a spirit of enduring hatred.

From that time forward, Mr. Utterson began to haunt the door in the by street of shops. In the morning before office hours, at noon when business was plenty of time scarce, at night under the face of the full city moon, by all lights and at all hours of solitude or concourse, the lawyer was to be found on his chosen post.

"If he be Mr. Hyde," he had thought, "I should be Mr. Seek."

7. What is the purpose of the use of repetition in the following passage?

   It was a night of little ease to his toiling mind, toiling in mere darkness and besieged by questions.

   a. It serves as a demonstration of the mental state of Mr. Lanyon.
   b. It is reminiscent of the church bells that are mentioned in the story.
   c. It mimics Mr. Utterson's ambivalence.
   d. It emphasizes Mr. Utterson's anguish in failing to identify Hyde's whereabouts.

8. What is the setting of the story in this passage?
   a. In the city
   b. On the countryside
   c. In a jail
   d. In a mental health facility

9. What can one infer about the meaning of the word "Juggernaut" from the author's use of it in the passage?
   a. It is an apparition that appears at daybreak.
   b. It scares children.
   c. It is associated with space travel.
   d. Mr. Utterson finds it soothing.

10. What is the definition of the word *haunt* in the following passage?

    From that time forward, Mr. Utterson began to haunt the door in the by street of shops. In the morning before office hours, at noon when business was plenty of time scarce, at night under the face of the full city moon, by all lights and at all hours of solitude or concourse, the lawyer was to be found on his chosen post.

   a. To levitate
   b. To constantly visit
   c. To terrorize
   d. To daunt

11. The phrase *labyrinths of lamplighted city* contains an example of what?
   a. Hyperbole
   b. Simile
   c. Juxtaposition
   d. Alliteration

12. What can one reasonably conclude from the final comment of this passage?

    "If he be Mr. Hyde," he had thought, "I should be Mr. Seek."

   a. The speaker is considering a name change.
   b. The speaker is experiencing an identity crisis.
   c. The speaker has mistakenly been looking for the wrong person.
   d. The speaker intends to continue to look for Hyde.

Questions 13–18 are based upon the following passage:

This excerpt is adaptation from "What to the Slave is the Fourth of July?" Rochester, New York, July 5, 1852.

> Fellow citizens—Pardon me, and allow me to ask, why am I called upon to speak here today? What have I, or those I represent, to do with your national independence? Are the great principles of political freedom and of natural justice embodied in that Declaration of Independence, Independence extended to us? And am I therefore called upon to bring our humble offering to the national altar, and to confess the benefits, and express devout gratitude for the blessings, resulting from your independence to us?
>
> Would to God, both for your sakes and ours, ours that an affirmative answer could be truthfully returned to these questions! Then would my task be light, and my burden easy and delightful. For who is there so cold that a nation's sympathy could not warm him? Who so obdurate and dead to the claims of gratitude that would not thankfully acknowledge such priceless benefits? Who so stolid and selfish, that would not give his voice to swell the hallelujahs of a nation's jubilee, when the chains of servitude had

been torn from his limbs? I am not that man. In a case like that, the dumb may eloquently speak, and the lame man leap as an hart.

But, such is not the state of the case. I say it with a sad sense of the disparity between us. I am not included within the pale of this glorious anniversary. Oh pity! Your high independence only reveals the immeasurable distance between us. The blessings in which you this day rejoice, I do not enjoy in common. The rich inheritance of justice, liberty, prosperity, and independence, bequeathed by your fathers, is shared by *you*, not by *me*. This Fourth of July is *yours,* not *mine*. You may rejoice, *I* must mourn. To drag a man in fetters into the grand illuminated temple of liberty, and call upon him to join you in joyous anthems, were inhuman mockery and sacrilegious irony. Do you mean, citizens, to mock me, by asking me to speak today? If so there is a parallel to your conduct. And let me warn you that it is dangerous to copy the example of a nation whose crimes, towering up to heaven, were thrown down by the breath of the Almighty, burying that nation and irrecoverable ruin! I can today take up the plaintive lament of a peeled and woe-smitten people.

By the rivers of Babylon, there we sat down. Yea! We wept when we remembered Zion. We hanged our harps upon the willows in the midst thereof. For there, they that carried us away captive, required of us a song; and they who wasted us required of us mirth, saying, "Sing us one of the songs of Zion." How can we sing the Lord's song in a strange land? If I forget thee, O Jerusalem, let my right hand forget her cunning. If I do not remember thee, let my tongue cleave to the roof of my mouth.

13. What is the tone of the first paragraph of this passage?
   a. Exasperated
   b. Inclusive
   c. Contemplative
   d. Nonchalant

14. Which word CANNOT be used synonymously with the term *obdurate* as it is conveyed in the text below?

   Who so obdurate and dead to the claims of gratitude, that would not thankfully acknowledge such priceless benefits?

   a. Steadfast
   b. Stubborn
   c. Contented
   d. Unwavering

15. What is the central purpose of this text?
   a. To demonstrate the author's extensive knowledge of the Bible
   b. To address the feelings of exclusion expressed by African Americans after the establishment of the Fourth of July holiday
   c. To convince wealthy landowners to adopt new holiday rituals
   d. To explain why minorities often relished the notion of segregation in government institutions

16. Which statement serves as evidence of the question above?
    a. By the rivers of Babylon . . . down.
    b. Fellow citizens . . . today.
    c. I can . . . woe-smitten people.
    d. The rich inheritance of justice . . . *not by me*.

17. The statement below features an example of which of the following literary devices?
    Oh pity! Your high independence only reveals the immeasurable distance between us.

    a. Assonance
    b. Parallelism
    c. Amplification
    d. Hyperbole

18. The speaker's use of biblical references, such as "rivers of Babylon" and the "songs of Zion," helps the reader to do all EXCEPT which of the following?
    a. Identify with the speaker through the use of common text.
    b. Convince the audience that injustices have been committed by referencing another group of people who have been previously affected by slavery.
    c. Display the equivocation of the speaker and those that he represents.
    d. Appeal to the listener's sense of humanity.

Questions 19–24 are based upon the following passage:

This excerpt is an adaptation from Abraham Lincoln's Address Delivered at the Dedication of the Cemetery at Gettysburg, November 19, 1863.

> Four score and seven years ago our fathers brought forth on this continent, a new nation, conceived in liberty, and dedicated to the proposition that all men are created equal.
>
> Now we are engaged in a great civil war, testing whether that nation, or any nation so conceived and so dedicated, can long endure. We are met on a great battlefield of that war. We have come to dedicate a portion of that field, as a final resting place for those who here gave their lives that this nation might live. It is altogether fitting and proper that we should do this.
>
> But, in a larger sense, we cannot dedicate—we cannot consecrate that we cannot hallow—this ground. The brave men, living and dead, who struggled here, have consecrated it, far above our poor power to add or detract. The world will little note, nor long remember what we say here, but it can never forget what they did here. It is for us the living, rather, to be dedicated here to the unfinished work which they who fought here have thus far so nobly advanced. It is rather for us to be here and dedicated to the great task remaining before us—that from these honored dead we take increased devotion to that cause for which they gave the last full measure of devotion—that we here highly resolve that these dead shall not have died in vain—that these this nation, under God, shall have a new birth of freedom—and that government of people, by the people, for the people, shall not perish from the earth.

19. The best description for the phrase *four score and seven years ago* is which of the following?
    a. A unit of measurement
    b. A period of time
    c. A literary movement
    d. A statement of political reform

20. What is the setting of this text?
    a. A battleship off of the coast of France
    b. A desert plain on the Sahara Desert
    c. A battlefield in North America
    d. The residence of Abraham Lincoln

21. Which war is Abraham Lincoln referring to in the following passage?

    > Now we are engaged in a great civil war, testing whether that nation, or any nation so conceived and so dedicated, can long endure.

    a. World War I
    b. The War of the Spanish Succession
    c. World War II
    d. The American Civil War

22. What message is the author trying to convey through this address?
    a. The audience should consider the death of the people that fought in the war as an example and perpetuate the ideals of freedom that the soldiers died fighting for.
    b. The audience should honor the dead by establishing an annual memorial service.
    c. The audience should form a militia that would overturn the current political structure.
    d. The audience should forget the lives that were lost and discredit the soldiers.

23. Which rhetorical device is being used in the following passage?

    > . . . we here highly resolve that these dead shall not have died in vain—that these this nation, under God, shall have a new birth of freedom—and that government of people, by the people, for the people, shall not perish from the earth.

    a. Antimetabole
    b. Antiphrasis
    c. Anaphora
    d. Epiphora

24. What is the effect of Lincoln's statement in the following passage?

    > But, in a larger sense, we cannot dedicate—we cannot consecrate that we cannot hallow—this ground. The brave men, living and dead, who struggled here, have consecrated it, far above our poor power to add or detract.

    a. His comparison emphasizes the great sacrifice of the soldiers who fought in the war.
    b. His comparison serves as a reminder of the inadequacies of his audience.
    c. His comparison serves as a catalyst for guilt and shame among audience members.
    d. His comparison attempts to illuminate the great differences between soldiers and civilians.

Questions 25–30 are based upon the following passage:

This excerpt is adaptation from Charles Dickens' speech in Birmingham in England on December 30, 1853 on behalf of the Birmingham and Midland Institute.

My Good Friends,—When I first imparted to the committee of the projected Institute my particular wish that on one of the evenings of my readings here the main body of my audience should be composed of working men and their families, I was animated by two desires; first, by the wish to have the great pleasure of meeting you face to face at this Christmas time, and accompany you myself through one of my little Christmas books; and second, by the wish to have an opportunity of stating publicly in your presence, and in the presence of the committee, my earnest hope that the Institute will, from the beginning, recognise one great principle—strong in reason and justice—which I believe to be essential to the very life of such an Institution. It is, that the working man shall, from the first unto the last, have a share in the management of an Institution which is designed for his benefit, and which calls itself by his name.

I have no fear here of being misunderstood—of being supposed to mean too much in this. If there ever was a time when any one class could of itself do much for its own good, and for the welfare of society—which I greatly doubt—that time is unquestionably past. It is in the fusion of different classes, without confusion; in the bringing together of employers and employed; in the creating of a better common understanding among those whose interests are identical, who depend upon each other, who are vitally essential to each other, and who never can be in unnatural antagonism without deplorable results, that one of the chief principles of a Mechanics' Institution should consist. In this world a great deal of the bitterness among us arises from an imperfect understanding of one another. Erect in Birmingham a great Educational Institution, properly educational; educational of the feelings as well as of the reason; to which all orders of Birmingham men contribute; in which all orders of Birmingham men meet; wherein all orders of Birmingham men are faithfully represented—and you will erect a Temple of Concord here which will be a model edifice to the whole of England.

Contemplating as I do the existence of the Artisans' Committee, which not long ago considered the establishment of the Institute so sensibly, and supported it so heartily, I earnestly entreat the gentlemen—earnest I know in the good work, and who are now among us,—by all means to avoid the great shortcoming of similar institutions; and in asking the working man for his confidence, to set him the great example and give him theirs in return. You will judge for yourselves if I promise too much for the working man, when I say that he will stand by such an enterprise with the utmost of his patience, his perseverance, sense, and support; that I am sure he will need no charitable aid or condescending patronage; but will readily and cheerfully pay for the advantages which it confers; that he will prepare himself in individual cases where he feels that the adverse circumstances around him have rendered it necessary; in a word, that he will feel his responsibility like an honest man, and will most honestly and manfully discharge it. I now proceed to the pleasant task to which I assure you I have looked forward for a long time.

25. Which word is most closely synonymous with the word *patronage* as it appears in the following statement?

> . . . that I am sure he will need no charitable aid or condescending patronage

a. Auspices
b. Aberration
c . Acerbic
d. Adulation

26. Which term is most closely aligned with the definition of the term *working man* as it is defined in the following passage?

> You will judge for yourselves if I promise too much for the working man, when I say that he will stand by such an enterprise with the utmost of his patience, his perseverance, sense, and support . . .

a. Plebeian
b. Viscount
c. Entrepreneur
d. Bourgeois

27. Which of the following statements most closely correlates with the definition of the term *working man* as it is defined in Question 26?

a. A working man is not someone who works for institutions or corporations, but someone who is well versed in the workings of the soul.
b. A working man is someone who is probably not involved in social activities because the physical demand for work is too high.
c. A working man is someone who works for wages among the middle class.
d. The working man has historically taken to the field, to the factory, and now to the screen.

28. Based upon the contextual evidence provided in the passage above, what is the meaning of the term *enterprise* in the third paragraph?

a. Company
b. Courage
c. Game
d. Cause

29. The speaker addresses his audience as *My Good Friends*—what kind of credibility does this salutation give to the speaker?

a. The speaker is an employer addressing his employees, so the salutation is a way for the boss to bridge the gap between himself and his employees.
b. The speaker's salutation is one from an entertainer to his audience and uses the friendly language to connect to his audience before a serious speech.
c. The salutation gives the serious speech that follows a somber tone, as it is used ironically.
d. The speech is one from a politician to the public, so the salutation is used to grab the audience's attention.

30. According to the aforementioned passage, what is the speaker's second desire for his time in front of the audience?

a. To read a Christmas story
b. For the working man to have a say in his institution which is designed for his benefit
c. To have an opportunity to stand in their presence
d. For the life of the institution to be essential to the audience as a whole

Questions 31–36 are based upon the following passage:

This excerpt is adaptation from *Our Vanishing Wildlife,* by William T. Hornaday

> Three years ago, I think there were not many bird-lovers in the United States, who believed it possible to prevent the total extinction of both egrets from our fauna. All the known rookeries accessible to plume-hunters had been totally destroyed. Two years ago, the secret discovery of several small, hidden colonies prompted William Dutcher, President of the National Association of Audubon Societies, and Mr. T. Gilbert Pearson, Secretary, to attempt the protection of those colonies. With a fund contributed for the purpose, wardens were hired and duly commissioned. As previously stated, one of those wardens was shot dead in cold blood by a plume hunter. The task of guarding swamp rookeries from the attacks of money-hungry desperadoes to whom the accursed plumes were worth their weight in gold, is a very chancy proceeding. There is now one warden in Florida who says that "before they get my rookery they will first have to get me."
>
> Thus far the protective work of the Audubon Association has been successful. Now there are twenty colonies, which contain all told, about 5,000 egrets and about 120,000 herons and ibises which are guarded by the Audubon wardens. One of the most important is on Bird Island, a mile out in Orange Lake, central Florida, and it is ably defended by Oscar E. Baynard. To-day, the plume hunters who do not dare to raid the guarded rookeries are trying to study out the lines of flight of the birds, to and from their feeding-grounds, and shoot them in transit. Their motto is—"Anything to beat the law, and get the plumes." It is there that the state of Florida should take part in the war.
>
> The success of this campaign is attested by the fact that last year a number of egrets were seen in eastern Massachusetts—for the first time in many years. And so to-day the question is, can the wardens continue to hold the plume-hunters at bay?

31. The author's use of first person pronoun in the following text does NOT have which of the following effects?

> Three years ago, I think there were not many bird-lovers in the United States, who believed it possible to prevent the total extinction of both egrets from our fauna.

a. The phrase *I think* acts as a sort of hedging, where the author's tone is less direct and/or absolute.
b. It allows the reader to more easily connect with the author.
c. It encourages the reader to empathize with the egrets.
d. It distances the reader from the text by overemphasizing the story.

32. What purpose does the quote serve at the end of the first paragraph?
    a. The quote shows proof of a hunter threatening one of the wardens.
    b. The quote lightens the mood by illustrating the colloquial language of the region.
    c. The quote provides an example of a warden protecting one of the colonies.
    d. The quote provides much needed comic relief in the form of a joke.

33. What is the meaning of the word *rookeries* in the following text?

    To-day, the plume hunters who do not dare to raid the guarded rookeries are trying to study out the lines of flight of the birds, to and from their feeding-grounds, and shoot them in transit.

    a. Houses in a slum area
    b. A place where hunters gather to trade tools
    c. A place where wardens go to trade stories
    d. A colony of breeding birds

34. What is on Bird Island?
    a. Hunters selling plumes
    b. An important bird colony
    c. Bird Island Battle between the hunters and the wardens
    d. An important egret with unique plumes

35. What is the main purpose of the passage?
    a. To persuade the audience to act in preservation of the bird colonies
    b. To show the effect hunting egrets has had on the environment
    c. To argue that the preservation of bird colonies has had a negative impact on the environment.
    d. To demonstrate the success of the protective work of the Audubon Association

36. Why are hunters trying to study the lines of flight of the birds?
    a. To study ornithology, one must know the lines of flight that birds take.
    b. To help wardens preserve the lives of the birds
    c. To have a better opportunity to hunt the birds
    d. To builds their homes under the lines of flight because they believe it brings good luck

Questions 37–42 are based upon the following passage:

This excerpt is adaptation from *The Life-Story of Insects,* by Geo H. Carpenter.

> Insects as a whole are preeminently creatures of the land and the air. This is shown not only by the possession of wings by a vast majority of the class, but by the mode of breathing to which reference has already been made, a system of branching air-tubes carrying atmospheric air with its combustion-supporting oxygen to all the insect's tissues. The air gains access to these tubes through a number of paired air-holes or spiracles, arranged segmentally in series.
>
> It is of great interest to find that, nevertheless, a number of insects spend much of their time under water. This is true of not a few in the perfect winged state, as for example aquatic beetles and water-bugs ('boatmen' and 'scorpions') which have some way of protecting their spiracles when submerged, and, possessing usually the power of flight, can pass on occasion from pond or stream to upper air. But it is advisable in connection with our present subject to dwell especially on some insects that remain continually

under water till they are ready to undergo their final moult and attain the winged state, which they pass entirely in the air. The preparatory instars of such insects are aquatic; the adult instar is aerial. All may-flies, dragon-flies, and caddis-flies, many beetles and two-winged flies, and a few moths thus divide their life-story between the water and the air. For the present we confine attention to the Stone-flies, the May-flies, and the Dragon-flies, three well-known orders of insects respectively called by systematists the Plecoptera, the Ephemeroptera and the Odonata.

In the case of many insects that have aquatic larvae, the latter are provided with some arrangement for enabling them to reach atmospheric air through the surface-film of the water. But the larva of a stone-fly, a dragon-fly, or a may-fly is adapted more completely than these for aquatic life; it can, by means of gills of some kind, breathe the air dissolved in water.

37. Which statement best details the central idea in this passage?
   a. It introduces certain insects that transition from water to air.
   b. It delves into entomology, especially where gills are concerned.
   c. It defines what constitutes as insects' breathing.
   d. It invites readers to have a hand in the preservation of insects.

38. Which definition most closely relates to the usage of the word *moult* in the passage?
   a. An adventure of sorts, especially underwater
   b. Mating act between two insects
   c. The act of shedding part or all of the outer shell
   d. Death of an organism that ends in a revival of life

39. What is the purpose of the first paragraph in relation to the second paragraph?
   a. The first paragraph serves as a cause and the second paragraph serves as an effect.
   b. The first paragraph serves as a contrast to the second.
   c. The first paragraph is a description for the argument in the second paragraph.
   d. The first and second paragraphs are merely presented in a sequence.

40. What does the following sentence most nearly mean?

   The preparatory instars of such insects are aquatic; the adult instar is aerial.

   a. The volume of water is necessary to prep the insect for transition rather than the volume of the air.
   b. The abdomen of the insect is designed like a star in the water as well as the air.
   c. The stage of preparation in between molting is acted out in the water, while the last stage is in the air.
   d. These insects breathe first in the water through gills, yet continue to use the same organs to breathe in the air.

Answer Explanations

1. C: Gulliver becomes acquainted with the people and practices of his new surroundings. Choice *C* is the correct answer because it most extensively summarizes the entire passage. While Choices *A* and *B* are reasonable possibilities, they reference portions of Gulliver's experiences, not the whole. Choice *D* is incorrect because Gulliver doesn't express repentance or sorrow in this particular passage.

2. A: Principal refers to *chief* or *primary* within the context of this text. Choice *A* is the answer that most closely aligns with this answer. Choices *B* and *D* make reference to a helper or followers while Choice *C* doesn't meet the description of Gulliver from the passage.

3. C: One can reasonably infer that Gulliver is considerably larger than the children who were playing around him because multiple children could fit into his hand. Choice *B* is incorrect because there is no indication of stress in Gulliver's tone. Choices *A* and *D* aren't the best answer because though Gulliver seems fond of his new acquaintances, he didn't travel there with the intentions of meeting new people or to express a definite love for them in this particular portion of the text.

4. C: The emperor made a *definitive decision* to expose Gulliver to their native customs. In this instance, the word *mind* was not related to a vote, question, or cognitive ability.

5. A: Choice *A* is correct. This assertion does *not* support the fact that games are a commonplace event in this culture because it mentions conduct, not games. Choices *B*, *C*, and *D* are incorrect because these do support the fact that games were a commonplace event.

6. B: Choice *B* is the only option that mentions the correlation between physical ability and leadership positions. Choices *A* and *D* are unrelated to physical strength and leadership abilities. Choice *C* does not make a deduction that would lead to the correct answer—it only comments upon the abilities of common townspeople.

7. D: It emphasizes Mr. Utterson's anguish in failing to identify Hyde's whereabouts. Context clues indicate that Choice *D* is correct because the passage provides great detail of Mr. Utterson's feelings about locating Hyde. Choice *A* does not fit because there is no mention of Mr. Lanyon's mental state. Choice *B* is incorrect; although the text does make mention of bells, Choice *B* is not the *best* answer overall. Choice *C* is incorrect because the passage clearly states that Mr. Utterson was determined, not unsure.

8. A: In the city. The word *city* appears in the passage several times, thus establishing the location for the reader.

9. B: It scares children. The passage states that the Juggernaut causes the children to scream. Choices *A* and *D* don't apply because the text doesn't mention either of these instances specifically. Choice *C* is incorrect because there is nothing in the text that mentions space travel.

10. B: To constantly visit. The mention of *morning*, *noon*, and *night* make it clear that the word *haunt* refers to frequent appearances at various locations. Choice *A* doesn't work because the text makes no mention of levitating. Choices *C* and *D* are not correct because the text makes mention of Mr. Utterson's anguish and disheartenment because of his failure to find Hyde but does not make mention of Mr. Utterson's feelings negatively affecting anyone else.

11. D: This is an example of alliteration. Choice *D* is the correct answer because of the repetition of the *L*-words. Hyperbole is an exaggeration, so Choice *A* doesn't work. No comparison is being made, so no simile or metaphor is being used, thus eliminating Choices *B* and *C*.

12. D: The speaker intends to continue to look for Hyde. Choices *A* and *B* are not possible answers because the text doesn't refer to any name changes or an identity crisis, despite Mr. Utterson's extreme obsession with finding Hyde. The text also makes no mention of a mistaken identity when referring to Hyde, so Choice *C* is also incorrect.

13. A: The tone is exasperated. While contemplative is an option because of the inquisitive nature of the text, Choice *A* is correct because the speaker is annoyed by the thought of being included when he felt that the fellow members of his race were being excluded. The speaker is not nonchalant, nor accepting of the circumstances which he describes.

14. C: Choice *C*, *contented*, is the only word that has different meaning. Furthermore, the speaker expresses objection and disdain throughout the entire text.

15. B: To address the feelings of exclusion expressed by African Americans after the establishment of the Fourth of July holiday. While the speaker makes biblical references, it is not the main focus of the passage, thus eliminating Choice *A* as an answer. The passage also makes no mention of wealthy landowners and doesn't speak of any positive response to the historical events, so Choices *C* and *D* are not correct.

16. D: Choice *D* is the correct answer because it clearly makes reference to justice being denied.

17. D: Hyperbole. Choices *A* and *B* are unrelated. Assonance is the repetition of sounds and commonly occurs in poetry. Parallelism refers to two statements that correlate in some manner. Choice *C* is incorrect because amplification normally refers to clarification of meaning by broadening the sentence structure, while hyperbole refers to a phrase or statement that is being exaggerated.

18. C: Choice *C* is correct because the speaker is clear about his intention and stance throughout the text; thus, it's not true that he makes biblical references to display his own equivocation and that of those that he represents. Choice *A* could be true, but the words "common text" is arguable because not everyone will understand the reference. Choice *B* is also partially true, as another group of people affected by slavery are being referenced. However, the speaker is not trying to convince the audience that injustices have been committed, as it is already understood there have been injustices committed. Choice *D* is also close to the correct answer, but it is not the best answer choice possible.

19. B: A period of time. It is apparent that Lincoln is referring to a period of time within the context of the passage because of how the sentence is structured with the word *ago*.

20. C: Lincoln's reference to *the brave men, living and dead, who struggled here,* proves that he is referring to a battlefield. Choices *A* and *B* are incorrect, as a *civil war* is mentioned and not a war with France or a war in the Sahara Desert. Choice *D* is incorrect because it does not make sense to consecrate a President's ground instead of a battlefield ground for soldiers who died during the American Civil War.

21. D: Abraham Lincoln is the former president of the United States, and he references a "civil war" during his address.

22. A: The audience should consider the death of the people that fought in the war as an example and perpetuate the ideals of freedom that the soldiers died fighting for. Lincoln doesn't address any of the topics outlined in Choices *B*, *C*, or *D*. Therefore, Choice *A* is the correct answer.

23. D: Choice *D* is the correct answer because of the repetition of the word *people* at the end of the passage. Choice A, antimetabole, is the repetition of words in a phrase or clause but in reverse order, such as: "I do what I like, and like what I do." Choice *B*, *antiphrasis*, is a form of denial of an assertion in a text. Choice *C*, *anaphora*, is the repetition that occurs at the beginning of sentences.

24. A: Choice *A* is correct because Lincoln's intention was to memorialize the soldiers who had fallen as a result of war as well as celebrate those who had put their lives in danger for the sake of their country. Choices *B* and *D* are incorrect because Lincoln's speech was supposed to foster a sense of pride among the members of the audience while connecting them to the soldiers' experiences.

25. A: The word *patronage* most nearly means *auspices*, which means *protection* or *support*. Choice *B*, *aberration*, means *deformity* and does not make sense within the context of the sentence. Choice *C*, *acerbic*, means *bitter* and also does not make sense in the sentence. Choice *D*, *adulation*, is a positive word meaning *praise*, and thus does not fit with the word *condescending* in the sentence.

26. D: *Working man* is most closely aligned with Choice *D*, *bourgeois*. In the context of the speech, the word *bourgeois* means *working* or *middle class*. Choice *A*, *Plebeian*, does suggest *common people*; however, this is a term that is specific to ancient Rome. Choice *B*, *viscount*, is a European title used to describe a specific degree of nobility. Choice *C*, *entrepreneur*, is a person who operates their own business.

27. C: In the context of the speech, the term *working man* most closely correlates with Choice *C*, *working man is someone who works for wages among the middle class.* Choice *A* is not mentioned in the passage and is off-topic. Choice *B* may be true in some cases, but it does not reflect the sentiment described for the term *working man* in the passage. Choice *D* may also be arguably true. However, it is not given as a definition but as *acts* of the working man, and the topics of *field, factory,* and *screen* are not mentioned in the passage.

28. D: *Enterprise* most closely means *cause*. Choices *A*, *B*, and *C* are all related to the term *enterprise*. However, Dickens speaks of a *cause* here, not a company, courage, or a game. *He will stand by such an enterprise* is a call to stand by a cause to enable the working man to have a certain autonomy over his own economic standing. The very first paragraph ends with the statement that the working man *shall . . . have a share in the management of an institution which is designed for his benefit.*

29. B: The speaker's salutation is one from an entertainer to his audience and uses the friendly language to connect to his audience before a serious speech. Recall in the first paragraph that the speaker is there to "accompany [the audience] . . . through one of my little Christmas books," making him an author there to entertain the crowd with his own writing. The speech preceding the reading is the passage itself, and, as the tone indicates, a serious speech addressing the "working man." Although the passage speaks of employers and employees, the speaker himself is not an employer of the audience, so Choice *A* is incorrect. Choice *C* is also incorrect, as the salutation is not used ironically, but sincerely, as the speech addresses the wellbeing of the crowd. Choice *D* is incorrect because the speech is not given by a politician, but by a writer.

30. B: For the working man to have a say in his institution which is designed for his benefit Choice *A* is incorrect because that is the speaker's *first* desire, not his second. Choices *C* and *D* are tricky because the language of both of these is mentioned after the word *second*. However, the speaker doesn't get to the second wish until the next sentence. Choices *C* and *D* are merely prepositions preparing for the statement of the main clause, Choice *B.*

31. D: The use of "I" could serve to have a "hedging" effect, allow the reader to connect with the author in a more personal way, and cause the reader to empathize more with the egrets. However, it doesn't distance the reader from the text, making Choice D the answer to this question.

32. C: The quote provides an example of a warden protecting one of the colonies. Choice *A* is incorrect because the speaker of the quote is a warden, not a hunter. Choice *B* is incorrect because the quote does not lighten the mood, but shows the danger of the situation between the wardens and the hunters. Choice *D* is incorrect because there is no humor found in the quote.

33. D: A *rookery* is a colony of breeding birds. Although *rookery* could mean Choice *A*, houses in a slum area, it does not make sense in this context. Choices *B* and *C* are both incorrect, as this is not a place for hunters to trade tools or for wardens to trade stories.

34. B: An important bird colony. The previous sentence is describing "twenty colonies" of birds, so what follows should be a bird colony. Choice *A* may be true, but we have no evidence of this in the text. Choice *C* does touch on the tension between the hunters and wardens, but there is no official "Bird Island Battle" mentioned in the text. Choice *D* does not exist in the text.

35. D: To demonstrate the success of the protective work of the Audubon Association. The text mentions several different times how and why the association has been successful and gives examples to back this fact. Choice *A* is incorrect because although the article, in some instances, calls certain people to act, it is not the purpose of the entire passage. There is no way to tell if Choices *B* and *C* are correct, as they are not mentioned in the text.

36. C: To have a better opportunity to hunt the birds. Choice *A* might be true in a general sense, but it is not relevant to the context of the text. Choice *B* is incorrect because the hunters are not studying lines of flight to help wardens, but to hunt birds. Choice *D* is incorrect because nothing in the text mentions that hunters are trying to build homes underneath lines of flight of birds for good luck.

37. A: It introduces certain insects that transition from water to air. Choice *B* is incorrect because although the passage talks about gills, it is not the central idea of the passage. Choices *C* and *D* are incorrect because the passage does not "define" or "invite," but only serves as an introduction to stoneflies, dragonflies, and mayflies and their transition from water to air.

38. C: The act of shedding part or all of the outer shell. Choices *A*, *B*, and *D* are incorrect.

39. B: The first paragraph serves as a contrast to the second. Notice how the first paragraph goes into detail describing how insects are able to breathe air. The second paragraph acts as a contrast to the first by stating "[i]t is of great interest to find that, nevertheless, a number of insects spend much of their time under water." Watch for transition words such as "nevertheless" to help find what type of passage you're dealing with.

40: C: The stage of preparation in between molting is acted out in the water, while the last stage is in the air. Choices *A, B,* and *D* are all incorrect. *Instars* is the phase between two periods of molting, and the text explains when these transitions occur.

Verbal

Vocabulary

Defining Words and English Origins

A **word** is a group of letters joined to form a single meaning. On their own, letters represent single sounds, but when placed together in a certain order, letters represent a specific image in the reader's mind in a way that provides meaning. Words can be nouns, verbs, adjectives, and adverbs, among others. Words also represent a verb tense of past, present, or future. Words allow for effective communication for commerce, social progress, technical advances, and much more. Simply put, words allow people to understand one another and create meaning in a complex world.

Throughout history, English words were shaped by other cultures and languages, such as Greek, Latin, French, Spanish, German, and others. They were borne from inventions, discoveries, and literary works, such as plays or science fiction novels. Others formed by shortening words that were already in existence. Some words evolved from the use of acronyms, such as *radar* (Radio Detection and Ranging) and *scuba* (Self-Contained Underwater Breathing Apparatus). The English language will continue to evolve as the needs and values of its speakers evolve.

Word Formation

How do English words form? They can be single root words, such as *love, hate, boy,* or *girl.* A **root word** is a word in its most basic form that carries a clear and distinct meaning. Complex English words combine affixes with root words. Some words have no root word, but are instead formed by combining various affixes, such as reject: *re-* is defined as repeating an action or actions, and *-ject* means throw or thrown. Therefore, **reject** is defined as the act of being thrown back. Words consisting of affixes alone are not the norm. Most words consist of either root words on their own or with the addition of affixes. To **affix** is to attach to something. Therefore, affixes in linguistic study are groups of letters that attach themselves to the beginning, middle, or end of root words to enhance or alter the word's meaning. Affixes added to verbs can change the word's tense, and affixes added to nouns can change the word's part of speech from noun to adjective, verb, or adverb.

Roots and Root Words

Words that exist on their own, without affixes, are root words. **Root words** are words written in their most basic form, and they carry a clear and distinct meaning. Consider the word *safe*. The root word, *safe*, acts as both a noun and adjective, and stands on its own, carrying a clear and distinct meaning.

The root of a word however, is not necessarily a part of the word that can stand on its own, although it does carry meaning. Since many English words come from Latin and Greek roots, it's helpful to have a general understanding of roots. Here is a list of common Greek and Latin roots used in the English language:

Root	Definition	Example
ami	Love	Amiable
ethno	Race	Ethnological
infra	Beneath or Below	Infrastructure
lun	Moon	Lunar
max	Greatest	Maximum
pent/penta	Five	Pentagon
sol	Sun	Solar
vac	Empty	Vacant

Affixes

Affixes are groups of letters that when added to the beginning or ending of root words, or are attachments within a root or root word itself, can:

- Intensify the word's meaning
- Create a new meaning
- Somewhat alter the existing meaning
- Change the verb tense
- Change the part of speech

There are three types of affixes: prefixes, suffixes, and infixes.

Prefixes

Prefixes are groups of letters attached to the beginning of a root word. *Pre-* refers to coming before, and *fix* refers to attaching to something. Consider the example of the root word *freeze*:

- Freeze: verb – to change from a liquid to solid by lowering the temperature to a freezing state.
- *Anti*freeze: noun – a liquid substance that prevents freezing when added to water, as in a vehicle's radiator.

By adding the prefix *anti-* to the root word *freeze*, the part of speech changed from verb to noun, and completely altered the meaning. *Anti-* as a prefix always creates the opposite in meaning, or the word's antonym.

By having a basic understanding of how prefixes work and what their functions are in a word's meaning, English speakers strengthen their fluency. Here is a list of some common prefixes in the English language, accompanied by their meanings:

Prefix	Definition	Example
ante-	before	antecedent
ex-	out/from	expel
inter-	between/among	intergalactic
multi-	much/many	multitude
post-	after	postscript
sub-	under	submarine
trans-	move between/across	transport
uni-	single/one	universe

Suffixes

Suffixes are groups of letters attached to the ending of a root or root word. Like prefixes, suffixes can:

- Intensify the word's meaning
- Create a new meaning
- Somewhat alter the existing meaning
- Change the verb tense
- Change the part of speech

Consider the example of the root word *fish* when suffixes are added:

- Fish: noun – a cold-blooded animal that lives completely in water and possesses fins and gills.
- Fishing: noun – I love the sport of fishing.
- Fishing: verb – Are you fishing today?

With the addition of the suffix *-ing*, the meaning of root word *fish* is altered, as is the part of speech.

A verb tense shift is made with the addition of the suffix *-ed*:

- Jump: present tense of to jump as in "I jump."
- Jumped: past tense of to jump as in "I jumped."
- Climb: present tense of to climb as in "We climb."
- Climbed: past tense of to climb as in "We climbed."

Here are a few common suffixes in the English language, along with their meanings:

Suffix	Meaning	Example
-ed	past tense	cooked
-ing	materials, present action	clothing
-ly	in a specific manner	lovely
-ness	a state or quality	brightness
-ment	action	enjoyment
-script	to write	transcript
-ee	receiver/performer	nominee
-ation/-ion	action or process	obligation

Infixes

Infixes are letters that attach themselves inside the root or root words. They generally appear in the middle of the word and are rare in the English language. Easily recognizable infixes include parent*s*-in-law, passer*s*-by, or cup*s*ful. Notice the ***-s*** is added inside the root word, making the word plural.

Special types of infixes, called **tmesis** words, are made from inserting an existing word into the middle of another word or between a compound word, creating a new word. Tmesis words are generally used in casual dialogue and slang speech. They add emphasis to the word's overall meaning and to evoke emotion on the part of the reader. Examples are fan-*bloody*-tastic and un-*freaking*-believable.

Tmesis words have been in existence since Shakespeare's time, as in this phrase from *Romeo and Juliet*, "…he is some *other*where." Shakespeare split up the compound word *somewhere* by inserting the word *other* between the two root words.

Compound Words

A **compound word** is created with the combination of two shorter words. To be a true compound word, two shorter words are combined, and the meaning of the longer word retains the meaning of the two shorter words. Compound words enhance the overall meaning, giving a broader description. There are three types of compound words: closed, hyphenated, and open.

Closed Compound Words

There was a time when closed compound words were not considered legitimate words. Over time and with continued, persistent use, they found a place in the English language. A **closed compound word** refers to a word that shows no separation between the two shorter words from which it is composed. Some examples of closed compound words are:

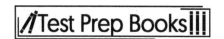

Closed Compound Word	Individual Words	Meaning
bookshelf	book/shelf	a shelf that holds books
doorstop	door/stop	an object to hold a door open
bedroom	bed/room	a room where one sleeps
bathroom	bath/room	a room where one bathes
backyard	back/yard	a yard in the back of a building
nightstand	night/stand	a small table beside one's bed

Hyphenated Compound Words

As the name suggests, **hyphenated compound words** include a hyphen that separates the two shorter words within the longer word. Some examples of hyphenated compound words are:

Hyphenated Compound Words	Individual Words	Meaning
self-service	self/serve	the act of serving one's self
color-blind	color/blind	incapable of accurately distinguishing colors
check-in	check/in	the act of registering as in attendance
year-round	year/round	any affair that takes place throughout the year
toll-free	toll/free	no application of toll/no charge
sugar-coated	sugar/coated	anything sweetened or coated with sugar

Open Compound Words

Open compound words appear as individual words but are they dependent on their partners to form the complete meaning of the compound word. Individually, the words may have meanings that are different than that of the pair together (the open compound word). For example, in *real estate, real* and *estate* have their own meanings that are different than the unique meaning of the compound word.

Open compound words are separated from each other by a single space and do not require a hyphen. Some examples of open compound words are:

Open Compound Words	**Meaning**
polka dot	a repeated circular dot that forms a pattern
sleeping bag	a special bag used to sleep in (usually when camping)
solar system	the collection of planets that orbit around the Sun
tape recorder	magnetic tape used to record sound
middle class	social group that is considered above lower class and below upper class
family room	a specific room for relaxation and entertainment used by all family members

Parts of Speech

Also referred to as word classes, **parts of speech** refer to the various categories in which words are placed. Words can be placed in any one or a combination of the following categories:

- Nouns
- Determiners
- Pronouns
- Verbs
- Adjectives
- Adverbs
- Prepositions
- Conjunctions

Understanding the various parts of speech used in the English language helps readers to better understand the written language.

Nouns

A **noun** is defined as any word that represents a person, place, animal, object, or idea. Nouns can identify a person's title or name, a person's gender, and a person's nationality, such as banker, commander-in-chief, female, male, or an American.

With animals, nouns identify the kingdom, phylum, class, etc. For example: the animal is an *elephant*, the phylum is *chordata*, and the class is *Mammalia*. It should be noted that the words *animal, phylum,* and *class* in the previous sentence are also nouns.

When identifying places, nouns refer to a physical location, a general vicinity, or the proper name of a city, state, or country. Some examples include the *desert*, the *East*, *Phoenix*, the *bathroom*, *Arizona*, an *office*, or the *United States*.

There are eight types of nouns: common, proper, countable, uncountable, concrete, abstract, compound, and collective.

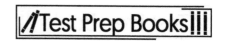

Common nouns are used in general terms, without specific identification. Examples include *girl, boy, country*, or *school*. Proper nouns refer to the specific proper name given to people, places, animals, or entities, such as *Melissa, Martin, Italy*, or *Harvard*.

Countable nouns can be counted: *one car, two cars*, or *three cars*. **Uncountable nouns** cannot be counted, such as *air, liquid*, or *gas*.

To be abstract is to exist, but only in thought or as an idea. An **abstract noun** cannot be physically touched, seen, smelled, heard, or tasted. Examples include *chivalry, day, fear, thought, truth, friendship*, or *freedom*.

To be **concrete** is to be seen, touched, tasted, heard, and/or smelled. Examples include *pie, snow, tree, bird, desk, hair*, or *dog*.

A **compound noun** is another term for an open compound word. Any noun that is written as two nouns that together form a specific meaning is a compound noun. For example, *post office, ice cream*, or *swimming pool*.

A **collective noun** refers to groups or collection of things that together form the whole. The members of the group are often people or individuals. Examples include *orchestra, squad, committee*, or the *majority*. It should be noted that the nouns in these examples are singular, but the word itself refers to a group containing more than one individual.

Determiners

Determiners modify a noun and usually refer to something specific. Determiners fall into one of four categories: *articles, demonstratives, quantifiers*, or *possessive determiners*.

Articles can be both definite articles, as in *the*, and indefinite as in *a, an, some*, or *any*:

> *The* man with *the* red hat.
>
> *A* flower growing in *the* yard.
>
> *Any* person who visits *the* store.

There are four different types of demonstratives: *this, that, these,* and *those*.

True demonstrative words will not directly precede the noun of the sentence but will be the noun. Some examples:

> *This* is the one.
>
> *That* is the place.
>
> *Those* are the files.

Once a demonstrative is placed directly in front of the noun, it becomes a demonstrative pronoun:

This one is perfect.

That place is lovely.

Those boys are annoying.

Quantifiers proceed nouns to give additional information to the noun about how much or how many is referred to. They can be used with countable and uncountable nouns:

She bought *plenty* of apples.

Few visitors came.

I got a *little* change.

Possessive determiners, sometimes called **possessive adjectives**, indicate possession. They are the possessive forms of personal pronouns, such as *for my, your, hers, his, its, their*, or *our*:

That is *my* car.

Tom rode *his* bike today.

Those papers are *hers.*

Pronouns

Pronouns are words that stand in place of nouns. There are three different types of pronouns: **subjective pronouns** (*I, you, he, she, it, we, they*), **objective pronouns** (*me, you, him, her it, us, them*), and **possessive pronouns** (*mine, yours, his, hers, ours, theirs*).

Note that some words are found in more than one pronoun category. See examples and clarifications are below:

You are going to the movies.

In the previous sentence, *you* is a subjective pronoun; it is the subject of the sentence and is performing the action.

I threw the ball to *you.*

Here, *you* is an objective pronoun; it is receiving the action and is the object of the sentence.

We saw *her* at the movies.

Her is an objective pronoun; it is receiving the action and is the object of the sentence.

The house across the street from the park is *hers.*

In this example, *hers* is a possessive pronoun; it shows possession of the house.

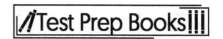

Verbs

Verbs are words in a sentence that show action or state. A sentence must contain a subject and a verb. Without a verb, a sentence is incomplete. Verbs may be in the present, past, or future tenses. To form auxiliary, or helping, verbs are required for some tenses in the future and in the past.

> I *see* the neighbors across the street.

See is an action.

> We *were eating* at the picnic.

Eating is the main action, and the verb *were* is the past tense of the verb *to be*, and is the helping or auxiliary verb that places the sentence in the past tense.

> You *will turn* 20 years of age next month.

Turn is the main verb, but *will* is the helping verb to show future tense of the verb *to be*.

Adjectives

Adjectives are a special group of words used to modify or describe a noun. Adjectives provide more information about the noun they modify. For example:

> The boy went to school. (There is no adjective.)

Rewriting the sentence, adding an adjective to further describe the boy and/or the school yields:

> The *young* boy went to the *old* school. (The adjective *young* describes the boy, and the adjective *old* describes the school.)

Adverbs

Adverb can play one of two roles: to modify the adjective or to modify the verb. For example:

> The young boy went to the old school.

We can further describe the adjectives *young* and *old* with adverbs placed directly in front of the adjectives:

> The *very* young boy went to the *very* old school. (The adverb *very* further describes the adjectives *young* and *old*.)

Other examples of using adverbs to further describe verbs include:

> The boy *slowly* went to school.

> The boy *happily* went to school.

The adverbs *slowly* and *happily* further modify the verbs.

Prepositions

Prepositions are special words that generally precede a noun. Prepositions clarify the relationship between the subject and another word or element in the sentence. They clarify time, place, and the positioning of the subjects and objects in a sentence. Common prepositions in the English language include: *near, far, under, over, on, in, between, beside, of, at, until, behind, across, after, before, for, from, to, by,* and *with*.

Conjunctions

Conjunctions are a group of unique words that connect clauses or sentences. They also work to coordinate words in the same clause. It is important to choose an appropriate conjunction based on the meaning of the sentence. Consider these sentences:

> I really like the flowers, *however* the smell is atrocious.
>
> I really like the flowers, *besides* the smell is atrocious.

The conjunctions *however* and *besides* act as conjunctions, connecting the two ideas: *I really like the flowers,* and, *the smell is atrocious*. In the second sentence, the conjunction *besides* makes no sense and would confuse the reader. Conjunctions must be chosen that clearly state the intended message without ambiguity.

Some conjunctions introduce an opposing opinion, thought, or fact. They can also reinforce an opinion, introduce an explanation, reinforce cause and effect, or indicate time. For example:

> **Opposition**: She wishes to go to the movies, *but* she doesn't have the money.
>
> **Cause and effect**: The professor became ill, *so* the class was postponed.
>
> **Time**: They visited Europe *before* winter came.

Each conjunction serves a specific purpose in uniting two separate ideas. Below are common conjunctions in the English language:

Opposition	Cause & Effect	Reinforcement	Time	Explanation
however	Therefore	besides	afterward	for example
nevertheless	as a result	anyway	before	in other words
but	because of this	after all	firstly	for instance
although	Consequently	furthermore	next	such as

Synonyms

Synonyms are words that mean the same or nearly the same thing in the same language. When presented with several words and asked to choose the synonym, more than one word may be similar to the original. However, one word is generally the strongest match. Synonyms should always share the same part of speech. For instance, *shy* and *timid* are both adjectives and hold similar meanings. The words *shy* and *loner* are similar, but *shy* is an adjective while *loner* is a noun. Another way to test for the best synonym is to reread the sentence with each possible word and determine which one makes the most sense. Consider the following sentence: *He will love you forever.*

Now consider the words: *adore, sweet, kind,* and *nice*. They seem similar, but when used in the following applications with the initial sentence, not all of them are synonyms for *love.*

He will *adore* you forever.

He will *sweet* you forever.

He will *kind* you forever.

He will *nice* you forever.

In the first sentence, the word *love* is used as a verb. The best synonym from the list that shares the same part of speech is *adore*. *Adore* is a verb, and when substituted in the sentence, it is the only substitution that makes grammatical and semantic sense.

Synonyms can be found for nouns, adjectives, verbs, adverbs, and prepositions. Here are some examples of synonyms from different parts of speech:

- **Nouns**: clothes, wardrobe, attire, apparel
- **Verbs**: run, sprint, dash
- **Adjectives**: fast, quick, rapid, swift
- **Adverbs**: slowly, nonchalantly, leisurely
- **Prepositions**: near, proximal, neighboring, close

Here are several more examples of synonyms in the English language:

Word	Synonym	Meaning
smart	intelligent	having or showing a high level of intelligence
exact	specific	clearly identified
almost	nearly	not quite but very close
to annoy	to bother	to irritate
to answer	to reply	to form a written or verbal response
building	edifice	a structure that stands on its own with a roof and four walls
business	commerce	the act of purchasing, negotiating, trading, and selling
defective	faulty	when a device is not working or not working well

Antonyms

Antonyms are words that are complete opposites. As with synonyms, there may be several words that represent the opposite meaning of the word in question. When choosing an antonym, one should choose the word that represents as close to the exact opposite in meaning as the given word, and ensure it shares the same part of speech.

Here are some examples of antonyms:

- Nouns: predator – prey
- Verbs: love – hate
- Adjectives: good – bad
- Adverbs: slowly – swiftly
- Prepositions: above – below

Homonyms

Homonyms are words that sound alike but carry different meanings. There are two different types of homonyms: homophones and homographs.

Homophones

Homophones are words that sound alike but carry different meanings and spellings. In the English language, there are several examples of homophones. Consider the following list:

Word	Meaning	Homophone	Meaning
I'll	I + will	aisle	a specific lane between seats
Allowed	past tense of the verb, 'to allow'	aloud	to utter a sound out loud
Eye	a part of the body used for seeing	I	first-person singular
Ate	the past tense of the verb, 'to eat'	eight	the number preceding the number nine
Peace	the opposite of war	piece	part of a whole
Seas	large bodies of natural water	seize	to take ahold of/to capture

Homographs

Homographs are words that share the same spelling but carry different meanings and different pronunciations. Consider the following list:

Word	Meaning	Homograph	Meaning
bass	fish	bass	musical instrument
bow	a weapon used to fire arrows	bow	to bend
Polish	of or from Poland	polish	a type of shine (n); to shine (v)
desert	dry, arid land	desert	to abandon

Verbal Analogies

The verbal analogies test portion tests the candidate's ability to analyze words carefully and find connections in definition and/or context. The test taker must compare a selected set of words with answer choices and select the ideal word to complete the sequence. While these exercises draw upon knowledge of vocabulary, this is also a test of critical thinking and reasoning abilities. Naturally, such skills are critical for building a career. Mastering verbal analogies will help people think objectively, discern critical details, and communicate more efficiently.

Question Layout

Verbal analogy sections are on other standardized tests such as the SAT. The format remains basically the same. First, two words are paired together that provide a frame for the analogy, and then there is a third word that must be found as a match in kind. It may help to think of it like this: A is to B as C is to D. Examine the breakdown below:

Apple (A) is to fruit (B) as carrot (C) is to vegetable (D).

As shown above, there are four words: the first three are given and the fourth word is the answer that must be found. The first two words are given to set up the kind of analogy that is to be replicated for the next pair. We see that apple is paired with fruit. In the first pair, a specific food item, apple, is paired to the food group category it corresponds with, which is fruit. When presented with the third word in the verbal analogy, carrot, a word must be found that best matches carrot in the way that fruit matched with apple. Again, carrot is a specific food item, so a match should be found with the appropriate food group: vegetable! Here's a sample prompt:

Morbid is to dead as jovial is to

a. Hate.
b. Fear.
c. Disgust.
d. Happiness.
e. Desperation.

As with the apple and carrot example, here is an analogy frame in the first two words: morbid and dead. Again, this will dictate how the next two words will correlate with one another. The definition of *morbid* is: described as or appealing to an abnormal and unhealthy interest in disturbing and unpleasant subjects, particularly death and disease. In other words, *morbid* can mean ghastly or death-like, which is why the word *dead* is paired with it. *Dead* relates to morbid because it describes morbid. With this in mind, *jovial* becomes the focus. *Jovial* means joyful, so out of all the choices given, the closest answer describing jovial is *happiness* (D).

Prompts on the exam will be structured just like the one above. "A is to B as C is to ?" will be given, where the answer completes the second pair. Or sometimes, "A is to B as ? is to ?" is given, where the second pair of words must be found that replicate the relationship between the first pair. The only things that will change are the words and the relationships between the words provided.

Discerning the Correct Answer

While it wouldn't hurt in test preparation to expand vocabulary, verbal analogies are all about delving into the words themselves and finding the right connection, the right word that will fit an analogy. People preparing for the test shouldn't think of themselves as human dictionaries, but rather as detectives. Remember, how the first two words are connected dictates the second pair. From there, picking the correct answer or simply eliminating the ones that aren't correct is the best strategy.

Just like a detective, a test taker needs to carefully examine the first two words of the analogy for clues. It's good to get in the habit of asking the questions: What do the two words have in common? What makes them related or unrelated? How can a similar relationship be replicated with the word I'm given and the answer choices? Here's another example:

Pillage is to steal as meander is to

a. Stroll.
b. Burgle.
c. Cascade.
d. Accelerate.
e. Pinnacle.

Why is *pillage* paired with *steal*? In this example, *pillage* and *steal* are synonymous: they both refer to the act of stealing. This means that the answer is a word that means the same as *meander*, which is *stroll*. In this case, the defining relationship in the whole analogy was a similar definition.

However, what if test takers don't know what *stroll* or *meander* mean? Using logic helps to eliminate choices and pick the correct answer. Looking closer into the contexts of the words *pillage* and *steal*, here are a few facts: these are things that humans do; and while they are actions, these are not necessarily types of movement. Again, pick a word that will not only match the given word, but best completes the relationship. It wouldn't make sense that *burgle* (B) would be the correct choice because *meander* doesn't have anything to do with stealing, so that eliminates *burgle*. *Pinnacle* (E) also can be eliminated because this is not an action at all but a position or point of reference. *Cascade* (C) refers to pouring or falling, usually in the context of a waterfall and not in reference to people, which means we can eliminate *cascade* as well. While people do accelerate when they move, they usually do so under additional circumstances: they accelerate while running or driving a car. All three of the words we see in the analogy are actions that can be done independently of other factors. Therefore, *accelerate* (D) can be eliminated, and *stroll* (A) should be chosen. *Stroll* and *meander* both refer to walking or wandering, so this fits perfectly.

The process of elimination will help rule out wrong answers. However, the best way to find the correct answer is simply to differentiate the correct answer from the other choices. For this, test takers should go back to asking questions, starting with the chief question: What's the connection? There are actually many ways that connections can be found between words. The trick is to look for the answer that is consistent with the relationship between the words given. What is the prevailing connection? Here are a few different ways verbal analogies can be formed.

Finding Connections in Word Analogies

Connections in Categories

One of the easiest ways to choose the correct answer in word analogies is simply to group words together. Ask if the words can be compartmentalized into distinct categories. Here are some examples:

Terrier is to dog as mystery is to
- a. Thriller.
- b. Murder.
- c. Detective.
- d. Novel.
- e. Investigation.

This one might have been a little confusing, but when looking at the first two words in the analogy, this is clearly one in which a category is the prevailing theme. Think about it: a *terrier* is a type of *dog.* While there are several breeds of dogs that can be categorized as a terrier, in the end, all terriers are still dogs. Therefore, *mystery* needs to be grouped into a category. *Murders, detectives*, and i*nvestigations* can all be involved in a mystery plot, but a *murder* (B), a *detective* (C), or an *investigation* (E) is not necessarily a mystery. A *thriller* (A) is a purely fictional concept, a kind of story or film, just like a mystery. A *thriller* can describe a mystery, but the same issue appears as the other choices. What about *novel* (D)? For one thing, it's distinct from all the other terms. A *novel* isn't a component of a mystery, but a mystery can be a type of novel. The relationship fits: a terrier is a type of dog, just like a mystery is a type of novel.

Synonym/Antonym

Some analogies are based on words meaning the same thing or expressing the same idea. Sometimes it's the complete opposite!

Marauder is to brigand as
- a. King is to peasant.
- b. Juice is to orange.
- c. Soldier is to warrior.
- d. Engine is to engineer.
- e. Paper is to photocopier.

Here, soldier is to warrior (C) is the correct answer. *Marauders* and *brigands* are both thieves. They are synonyms. The only pair of words that fits this analogy is *soldier* and *warrior* because both terms describe combatants who fight.

Cap is to shoe as jacket is to
- a. Ring.
- b. T-shirt.
- c. Vest.
- d. Glasses.
- e. Pants.

Opposites are at play here because a *cap* is worn on the head/top of the person, while a *shoe* is worn on the foot/bottom. A jacket is worn on top of the body too, so the opposite of *jacket* would be *pants* (E) because these are worn on the bottom of the body. Often the prompts on the test provide a synonym or antonym relationship. Just consider if the sets in the prompt reflect similarity or stark difference.

Parts of a Whole

Another thing to consider when first looking at an analogy prompt is whether the words presented come together in some way. Do they express parts of the same item? Does one word complete the other? Are they connected by process or function?

Tire is to car as

a. Wing is to bird.
b. Oar is to boat.
c. Box is to shelf.
d. Hat is to head.
e. Knife is to sheath.

We know that the *tire* fits onto the *car's* wheels and this is what enables the car to drive on roads. The *tire* is part of the *car.* This is the same relationship as oar is to boat (B). The *oars* are attached onto a *boat* and enable a person to move and navigate the boat on water. At first glance, wing is to bird (A) seems to fit too, since a *wing* is a part of a *bird* that enables it to move through the air. However, since a tire and car are not alive and transport people, oar and boat fit better because they are also not alive and they transport people. Subtle differences between answer choices should be found.

Other Relationships

There are a number of other relationships to look for when solving verbal analogies. Some relationships focus on one word being a **characteristic/NOT a characteristic** of the other word. Sometimes the first word is **the source/comprised of** the second word. Still, other words are related by their **location**. Some analogies have **sequential** relationships, and some are **cause/effect** relationships. There are analogies that show **creator/provider** relationships with the **creation/provision**. Another relationship might compare an **object with its function** or a **user with his or her tool**. An analogy may focus on a **change of grammar** or a **translation of language**. Finally, one word of an analogy may have a relationship to the other word in its **intensity**. The type of relationship between the first two words of the analogy should be determined before continuing to analyze the second set of words. One effective method of determining a relationship between two words is to form a comprehensible sentence using both words and then to plug the answer choices into the same sentence. For example, consider the following analogy: *Bicycle is to handlebars as car is to steering wheel.* A sentence could be formed that says: A bicycle navigates using its handlebars; therefore, a car navigates using its steering wheel. If the second sentence makes sense, then the correct relationship has likely been found. A sentence may be more complex depending on the relationship between the first two words in the analogy. An example of this may be: *food is to dishwasher as dirt is to carwash.* The formed sentence may be: A dishwasher cleans food off of dishes in the same way that a carwash cleans dirt off of a car.

Dealing with Multiple Connections

There are many other ways to draw connections between word sets. Several word choices might form an analogy that would fit the word set in your prompt. When this occurs, the analogy must be explored from multiple angles as, on occasion, multiple answer choices may appear to be correct. When this occurs, ask yourself: which one is an even closer match than the others? The framing word pair is another important point to consider. Can one or both words be interpreted as actions or ideas, or are they purely objects? Here's a question where words could have alternate meanings:

Hammer is to nail as saw is to

a. Electric.
b. Hack.
c. Cut.
d. Machete.
e. Groove.

Looking at the question above, it becomes clear that the topic of the analogy involves construction tools. *Hammers* and *nails* are used in concert, since the hammer is used to pound the nail. The logical first thing to do is to look for an object that a saw would be used on. Seeing that there is no such object among the answer choices, a test taker might begin to worry. After all, that seems to be the choice that would complete the analogy—but that doesn't mean it's the only choice that may fit. Encountering questions like this test one's ability to see multiple connections between words—don't get stuck thinking that words can only be connected in a single way. The first two words given can be verbs instead of just objects. To *hammer* means to hit or beat; oftentimes it refers to beating something into place. This is also what *nail* means when it is used as a verb. Here are the word choices that reveal the answer.

First, it's known that a saw, well, saws. It uses a steady motion to cut an object, and indeed to *saw* means to *cut*! *Cut* (C) is one of our answer choices, but the other options should be reviewed. While some tools are *electric* (a), the use of power in the tools listed in the analogy isn't a factor. Again, it's been established that these word choices are not tools in this context. Therefore, *machete* (D) is also ruled out because machete is also not a verb. Another important thing to consider is that while a *machete* is a tool that accomplishes a similar purpose as a *saw*, the *machete* is used in a slicing motion rather than a sawing/cutting motion. The verb that describes machete is *hack* (B), another choice that can be ruled out. A *machete* is used to hack at foliage. However, a *saw* does not hack. *Groove* (E) is just a term that has nothing to do with the other words, so this choice can be eliminated easily. This leaves *cut* (C), which confirms that this is the word needed to complete the analogy.

Synonyms

This portion of the exam is specifically constructed to test vocabulary skills and the ability to discern the best answer that matches the provided word. Unlike verbal analogies, which will test communication skills and problem-solving abilities along with vocabulary, synonym questions chiefly test vocabulary knowledge. While logic and reasoning come into play in this section, they are not as heavily emphasized as with the analogies. A prior knowledge of what the words mean is helpful in order to answer correctly. If the meaning of the words is unknown, that's fine, too; strategies should be used to rule out false answers and choose the correct ones. Here are some study strategies for an optimum performance.

Question Format

In contrast to the verbal analogies, synonym questions are very simple in construction. Instead of a comparison of words with an underlying connection, the prompt is just a single word. There are no special directions, alternate meanings, or analogies to work with. The objective is to analyze the given word and then choose the answer that means the same thing or is closest in meaning to the given word. Note the example below:

Blustery
   a. Hard
   b. Windy
   c. Mythical
   d. Stony
   e. Corresponding

All of the questions on the synonym portion will appear exactly like the above sample. This is generally the standard layout throughout other exams, so some test takers may already be familiar with the structure. The principle remains the same: at the top of the section, clear directions will be given to choose the answer that most precisely defines the given word. In this case, the answer is *windy* (B), since *windy* and *blustery* are synonymous.

Preparation

In truth, there is no set way to prepare for this portion of the exam that will guarantee a perfect score. This is simply because the words used on the test are unpredictable. There is no set list provided to study from. The definition of the provided word needs to be determined on the spot. This sounds challenging, but there are still ways to prepare mentally for this portion of the test. It may help to expand your vocabulary a little each day. Several resources are available, in books and online, that collect words and definitions that tend to show up frequently on standardized tests. Knowledge of words can increase the strength of your vocabulary.

Mindset is key. The meanings of challenging words can often be found by relying on the past experiences of the test taker to help deduce the correct answer. How? Well, test takers have been talking their entire lives—knowing words and how words work. It helps to have a positive mindset from the start. It's unlikely that all definitions of words will be known immediately, but the answer can still be found. There are aspects of words that are recognizable to help discern the correct answers and eliminate the incorrect ones. Below are some of the factors that contribute to word meanings.

Word Origins and Roots

Studying a foreign language in school, particularly Latin or any of the romance languages (Latin-influenced), is advantageous. As mentioned, English is a language highly influenced by Latin and Greek words. The roots of much of the English vocabulary have Latin origins; these roots can bind many words together and often allude to a shared definition.

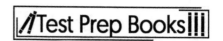

Here's an example:

Fervent
  a. Lame
  b. Joyful
  c. Thorough
  d. Boiling
  e. Cunning

Fervent descends from the Latin word, *fervere*, which means "to boil or glow" and figuratively means "impassioned." The Latin root present in the word is *ferv*, which is what gives fervent the definition: showing great warmth and spirit or spirited, hot, glowing. This provides a link to *boiling* (D) just by root word association, but there's more to analyze. Among the other choices, none relate to fervent. The word *lame* (A) means crippled, disabled, weak, or inadequate. None of these match with *fervent*. While being *fervent* can reflect joy, *joyful* (B) directly describes "a great state of happiness," while *fervent* is simply expressing the idea of having very strong feelings—not necessarily joy. *Thorough* (C) means complete, perfect, painstaking, or with mastery; while something can be done thoroughly and fervently, none of these words match *fervent* as closely as *boiling* does. *Cunning* (E) means crafty, deceiving or with ingenuity or dexterity. Doing something fervently does not necessarily mean it is done with dexterity. Not only does *boiling* connect in a linguistic way, but also in the way it is used in our language. While *boiling* can express being physically hot and undergoing a change, *boiling* is also used to reflect emotional states. People say they are "boiling over" when in heightened emotional states; "boiling mad" is another expression. *Boiling*, like *fervent*, also embodies a sense of heightened intensity. This makes *boiling* the best choice.

The Latin root *ferv* is seen in other words such as fervor, fervid, and even ferment. All of them are connected to and can be described by boil or glow, whether it is in a physical sense or in a metaphorical one. Such patterns can be seen in other word sets as well. Here's another example:

Gracious
  a. Fruitful
  b. Angry
  c. Grateful
  d. Understood
  e. Overheard

This one's a little easier; the answer is *grateful* (C), because both words mean thankful. Even if the meanings of both words are known, there's a connection found by looking at the beginnings of both words: *gra/grat*. Once again, these words are built on a root that stretches back to classical language. Both terms come from the Latin, *gratis*, which literally means "thanks."

Understanding root words can help identify the meaning in a lot of word choices, and help the test taker grasp the nature of the given word. Many dictionaries, both in book form and online, offer information on the origins of words, which highlight these roots. When studying for the test, it helps to look up an unfamiliar word for its definition and then check to see if it has a root that can be connected to any other terms.

Pay Attention to Prefixes

Recall that the prefix of a word can actually reveal a lot about its definition. Many prefixes are actually Greco-Roman roots as well—but these are more familiar and a lot easier to recognize! When encountering any unfamiliar words, try looking at prefixes to discern the definition and then compare that with the choices. The prefix should be determined to help find the word's meaning. Here's an example question:

Premeditate
- a. Sporadic
- b. Calculated
- c. Interfere
- d. Determined
- e. Noble

With *premeditate,* there's the common prefix *pre*. This helps draw connections to other words like prepare or preassemble. *Pre* refers to "before, already being, or having already." *Meditate* means to think or plan. *Premeditate* means to think or plan beforehand with intent. Therefore, a term that deals with thinking or planning should be found, but also something done in preparation. Among the word choices, *noble* (E) and *determined* (D) are both adjectives with no hint of being related to something done before or in preparation. These choices are wrong. *Sporadic* (A) refers to events happening in irregular patterns, so this is quite the opposite of premeditated. *Interfere* (C) also has nothing to do with premeditate; it goes counter to premeditate in a way similar to sporadic. *Calculated* (B), however, fits! A route and the cost of starting a plan can be calculated. Calculated refers to acting with a full awareness of consequences, so inherently planning is involved. In fact, calculated is synonymous with *premeditated*, thus making it the correct choice. Just by paying attention to a prefix, the doors to a meaning can open to help easily figure out which word would be the best choice. Here's another example.

Regain
- a. Erupt
- b. Ponder
- c. Seek
- d. Recoup
- e. Enamor

Recoup (D) is the right answer. The prefix *re* often appears in front of words to give them the meaning of occurring again. *Regain* means to repossess something that was lost. *Recoup*, which also has the *re* prefix, literally means to regain. In this example, both the given word and the answer share the *re* prefix, which makes the pair easy to connect. However, don't rely *only* on prefixes to choose an answer. Make sure to analyze all of the options before marking an answer. Going through the other words in this sample, none of them come close to meaning regain except recoup. After checking to make sure that recoup is the best matching word, then mark it!

Positive Versus Negative Sounding Words

Another tool for the mental toolbox is simply distinguishing whether a word has a positive or negative connotation. Like electrical wires, words carry energy; they are crafted to draw certain attention and to have certain strength to them. Words can be described as positive and uplifting (a stronger word) or they can be negative and scathing (a stronger word). Sometimes they are neutral—having no particular connotation. Distinguishing how a word is supposed to be interpreted will not only help learn its definition, but also draw parallels with word choices. While it's true that words must usually be taken in the context of how they are used, word definitions have inherent meanings as well, meaning that they have a distinct vibe to pick up on. Here is an example.

Excellent
  a. Fair
  b. Optimum
  c. Reasonable
  d. Negative
  e. Agitation

As you know, *excellent* is a very positive word. It refers to something being better than good, or above average. In this sample, *negative* (D) and *agitation* (E) can easily be eliminated because these are both words with negative connotations. *Reasonable* (C) is more or less a neutral word: it's not bad but it doesn't communicate the higher quality that excellent represents. It's just, well, reasonable. This leaves the possible choices of *fair* (A) and *optimum* (B). Or does it? Fair *is* a positive word; it's used to describe things that are good, even beautiful. But in the modern context, fair is defined as good, but somewhat average or just decent: "You did a fairly good job" or, "That was fair." On the other hand, optimum is positive and a stronger word. Optimum describes the most favorable outcome. This makes optimum the best word choice that matches excellent in both strength and connotation. Not only are the two words positive, but they also express the same level of positivity! Here's another sample.

Repulse
  a. Draw
  b. Encumber
  c. Force
  d. Disgust
  e. Magnify

Repulse just sounds negative when said aloud. It is commonly used in the context of something being repulsive, disgusting, or that which is distasteful. It's also defined as an attack that drives people away. This tells us that we need a word that also carries a negative meaning. *Magnify* (E) is positive, while *draw* (A) and *force* (C) are both neutral. *Encumber* (B) and *disgust* (D) are negative. *Disgust* is a stronger negative than *encumber*. Of all the words given, only *disgust* directly defines a feeling of distaste and aversion that is synonymous with *repulse* and matches in both negativity and strength.

Parts of Speech

It is often very helpful to determine the part of speech of a word. Is it an adjective, adverb, noun, or verb, etc.? Often the correct answer will also be the same part of speech as the given word. Isolate the part of speech and what it describes and look for an answer choice that also describes the same part of speech. For example: if the given word is an adverb describing an action word, then look for another adverb describing an action word.

Swiftly
  a. Fast
  b. Quietly
  c. Angry
  d. Sudden
  e. Quickly

Swiftly is an adverb that describes the speed of an action. *Angry* (C), *fast* (A), and *sudden* (D) can be eliminated because they are not adverbs, and *quietly* (B) can be eliminated because it does not describe speed. This leaves *quickly* (E), which is the correct answer. *Fast* and *sudden* may throw off some test takers because they both describe speed, but *quickly* matches more closely because it is an adverb, and *swiftly* is also an adverb.

Placing the Word in a Sentence

Often it is easier to discern the meaning of a word if it is used in a sentence. If the given word can be used in a sentence, then try replacing it with some of the answer choices to see which words seem to make sense in the same sentence. Here's an example.

Remarkable
  a. Often
  b. Capable
  c. Outstanding
  d. Shining
  e. Excluding

A sentence can be formed with the word *remarkable*. "My grade point average is remarkable." None of the examples make sense when replacing the word *remarkable* in the sentence other than the word *outstanding* (C), so *outstanding* is the obvious answer. *Shining* (D) is also a word with a positive connotation, but *outstanding* fits better in the sentence.

Looking for Relationships

Remember that all except one of the answer choices are wrong. If a close relationship between three or four of the answer choices can be found and not the fourth or fifth, then some of the choices can be eliminated. Sometimes all of the words are related except one; the one that is not related will often be the correct answer. Here is an example.

Outraged
  a. Angry
  b. Empty
  c. Forlorn
  d. Vacated
  e. Lonely

Notice that all of the answer choices have a negative connotation, but four of them are related to being alone or in low numbers. While two answer choices involve emotions—*angry* (A) and *lonely* (E), *lonely* is related to the other wrong answers, so *angry* is the best choice to match *outraged*.

Picking the Closest Answer

As the answer choices are reviewed, two scenarios might stand out. An exact definition match might not be found for the given word among the choices, or there are several word choices that can be considered synonymous to the given word. This is intentionally done to evaluate the test taker's ability to draw parallels between the words in order to produce an answer that best fits the prompt word. Again, the closest fitting word will be the answer. Even when facing these two circumstances, finding the one word that fits best is the proper strategy. Here's an example:

Insubordination

a. Cooperative
b. Disciplined
c. Rebel
d. Contagious
e. Wild

Insubordination refers to a defiance or utter refusal of authority. Looking over the choices, none of these terms provide definite matches to *insubordination* like insolence, mutiny, or misconduct would. This is fine; the answer doesn't have to be a perfect synonym. The choices don't reflect *insubordination* in any way, except rebel (C). After all, when *rebel* is used as a verb, it means to act against authority. It's also used as a noun: someone who goes against authority. Therefore, *rebel* is the best choice.

As with the verbal analogies section, playing the role of "detective" is the way to go as you may encounter two or even three answer choices that could be considered correct. However, the answer that best fits the prompt word's meaning is the best answer. Choices should be narrowed one word at a time. The least-connected word should be eliminated first and then proceed until one word is left that is the closest synonym.

Sequence

a. List
b. Range
c. Series
d. Replicate
e. Iconic

A *sequence* reflects a particular order in which events or objects follow. The two closest options are *list* (A) and *series* (C). Both involve grouping things together, but which fits better? Consider each word more carefully. A *list* is comprised of items that fit in the same category, but that's really it. A *list* doesn't have to follow any particular order; it's just a list. On the other hand, a *series* is defined by events happening in a set order. A *series* relies on sequence, and a *sequence* can be described as a series. Thus, *series* is the correct answer!

Practice Questions

Verbal Analogies

1. Begonia is to flower as
    a. Daisy is to pollen.
    b. Cat is to catnip.
    c. Cardiologist is to doctor.
    d. Radiology is to disease.
    e. Nutrition is to reproduction.

2. Malleable is to pliable as
    a. Corroborate is to invalidate.
    b. Avenue is to city.
    c. Blacksmith is to anvil.
    d. Hostile is to hospitable.
    e. Disparage is to criticize.

3. Cerebellum is to brain as
    a. Nurse is to medication.
    b. Nucleus is to cell.
    c. Bacteria is to spores.
    d. Refraction is to light.
    e. Painting is to sculpting.

4. Whisk is to baking as
    a. Glove is to boxing.
    b. Swimming is to water.
    c. Love is to romance.
    d. Azalea is to flower.
    e. Bench is to park.

5. Umpire is to officiate as
    a. Coaching is to coach.
    b. Baseball is to pastime.
    c. Counselor is to guide.
    d. Notary is to paper.
    e. Messenger is to letter.

6. Chuckle is to guffaw as
    a. Snicker is to lament.
    b. Affection is to nurturing.
    c. Sensitivity is to brevity.
    d. Whisper is to bellow.
    e. Serenity is to tranquility.

7. Fire is to passion as
   a. Sultry is to ferment.
   b. Emotion is to stagnant.
   c. Heat is to happiness.
   d. Ice is to rigidity.
   e. Comfort is to travel.

8. Geriatric is to youth as
   a. Transparent is to opaque.
   b. Soldier is to war.
   c. Subtle is to sophisticated.
   d. Yellow is to happiness.
   e. Surly is to scandalous.

9. Lying is to distrust as
   a. Prohibit is to outlaw.
   b. Petulant is to children.
   c. Exploit is to gain.
   d. Imprudent is to money.
   e. Hurricane is to devastation.

10. Desolate is to barren as
   a. Defend is to prosecute.
   b. Contend is to maintain.
   c. Avid is to cheerleader.
   d. Bittersweet is to happiness.
   e. King is to ambition.

11. Chapter is to novel as
   a. Whisper is to silence.
   b. Tangle is to lengthening.
   c. Poem is to poet.
   d. Feeling is to past.
   e. Stroke is to painting.

12. Carpenter is to construction as
   a. Adaptation is to insect.
   b. Acquisition is to possession.
   c. Wizard is to magic.
   d. Baker is to bread.
   e. Microsoft is to programming.

13. Tepid is to boiling as
   a. Moon is to femininity.
   b. Cornered is to immunity.
   c. Greedy is to rapacious.
   d. Sword is to dagger.
   e. Burning is to smoke.

14. Read is to learn as
    a. Ball is to soccer.
    b. Brand is to marketing.
    c. Exercise is to health.
    d. Disease is to energy.
    e. Party is to birthday.

15. Competent is to impotent as
    a. Demur is to accept.
    b. Dispute is to argument.
    c. Brandish is to gold.
    d. Honorary is to metal.
    e. Reach is to grab.

16. Merengue is to music as
    a. Cluster is to assemblage.
    b. Tension is to headache.
    c. Fiscal is to government.
    d. Hunger is to starving.
    e. Nomadic is to tribe.

17. Car is to transport as
    a. Radio is to sound.
    b. Volume is to voice.
    c. Triangles are to circles.
    d. Fireplace is to heat.
    e. Mangos are to fruit.

18. Flower is to femininity as
    a. Sunflower is to bees.
    b. Technology is to cell phones.
    c. Envy is to relationships.
    d. Plant is to carb dioxide.
    e. Light is to transcendence.

19. Principle is to truth as
    a. Squalid is to shabby.
    b. Frame is to picture.
    c. Static is to movement.
    d. Format is to index.
    e. Sour is to sweet.

20. Careful is to fastidious as
    a. Indulge is to deprive.
    b. Fluctuate is to trapeze.
    c. Majesty is to a lion.
    d. Endow is to bestow.
    e. Grieve is to lament.

21. Mantle is to earth as
    a. Volcano is to lava.
    b. Bundle is to uniform.
    c. Bun is to hamburger.
    d. Letter is to mailman.
    e. Spider is to spiderweb.

22. Maestro is to conducting as
    a. Barista is to coffee.
    b. Acupuncturist is to healing.
    c. Professor is to essay.
    d. President is to executive branch.
    e. Agent is to housing.

23. Contentious is to agreeable as
    a. Petulant is to irritable.
    b. Vocation is to career.
    c. Serendipity is to luck.
    d. Expedient is to useful.
    e. Penurious is to generous.

24. Cumulus is to cloud as
    a. Weather is to rain.
    b. Grape is to wine.
    c. Tortellini is to pasta.
    d. Elicit is to snow.
    e. Spanish is to English.

25. Pot is to boil as
    a. Belay is to climb.
    b. Water is to sink.
    c. Chef is to cook.
    d. Mirror is to reflection.
    e. Temper is to rage.

26. Laurel is to victory as
    a. Branch is to vine.
    b. Dove is to peace.
    c. Chair is to sit.
    d. Plumb is to apple.
    e. Curtain is to floor.

27. Transient is to ephemeral as
    a. Renounce is to acknowledge
    b. Placate is to subversive.
    c. Relinquish is to vapid.
    d. Sanguine is to cheerful.
    e. Verbose is to scanty.

28. Diverge is to agree as
    a. Assail is to belittle.
    b. Dire is to folly.
    c. Lurid is to spartan.
    d. Antipathy is to friendliness.
    e. Waning is to recant.

29. Nitrogen is to element as
    a. Species is to canine.
    b. Valley is to river.
    c. Crayon is to elementary.
    d. Project is to brand.
    e. Calico is to cat.

30. Small is to miniscule as
    a. Chagrin is to elusive.
    b. Confined is to rotten.
    c. Unhealthy is to ailing.
    d. Tall is to short.
    e. Cough is to allergy.

Synonyms

1. GARISH
    a. Drab
    b. Flashy
    c. Gait
    d. Hardy
    e. Lithe

2. INANE
    a. Ratify
    b. Illicit
    c. Uncouth
    d. Senseless
    e. Wry

3. SOLACE
    a. Marred
    b. Induce
    c. Depose
    d. Inherent
    e. Comfort

4. COPIOUS
   a. Dire
   b. Adept
   c. Indignant
   d. Ample
   e. Nuance

5. SUPERCILIOUS
   a. Tenuous
   b. Waning
   c. Arrogant
   d. Placate
   e. Extol

6. LURID
   a. Gruesome
   b. Placid
   c. Irate
   d. Quell
   e. Torpor

7. VANQUISH
   a. Saturate
   b. Conquer
   c. Reproach
   d. Parch
   e. Surrender

8. TRITE
   a. Scanty
   b. Banal
   c. Polemical
   d. Indulgent
   e. Eclectic

9. DIVULGE
   a. Dupe
   b. Flummox
   c. Indulgent
   d. Germinate
   e. Admit

10. INDOLENT
   a. Adamant
   b. Dour
   c. Noisome
   d. Lackadaisical
   e. Remiss

11. BOLSTER
    a. Bequeath
    b. Abate
    c. Support
    d. Palliate
    e. Tractable

12. UNWITTING
    a. Undermine
    b. Unintentional
    c. Rife
    d. Pernicious
    e. Stolid

13. UNGAINLY
    a. Clumsy
    b. Absurd
    c. Unruly
    d. Tenuous
    e. Petulant

14. PRATTLE
    a. Babble
    b. Prosaic
    c. Deluded
    d. Meddle
    e. Folly

15. PROLIFIC
    a. Devoid
    b. Elusive
    c. Laconic
    d. Productive
    e. Judicious

16. FORTITUDE
    a. Aura
    b. Disparage
    c. Finesse
    d. Cowardice
    e. Courage

17. ACUMEN
    a. Diligent
    b. Ingenuity
    c. Congenial
    d. Embroiled
    e. Reverent

18. RELEGATE
    a. Relay
    b. Temporize
    c. Demote
    d. Vigilant
    e. Spurn

19. CURTAIL
    a. Covet
    b. Abridge
    c. Foil
    d. Construe
    e. Bilk

20. CONSERVE
    a. Constrain
    b. Adjourn
    c. Stipulate
    d. Maintain
    e. Improve

21. AMBIVALENT
    a. Accosted
    b. Engrossed
    c. Impartial
    d. Conflicted
    e. Convinced

22. DISREPUTE
    a. Benevolent
    b. Condone
    c. Dishonor
    d. Emit
    e. Postulate

23. INSTIGATE
    a. Provoke
    b. Renounce
    c. Prescribe
    d. Modify
    e. Diminish

24. MAXIM
    a. Discord
    b. Clout
    c. Apex
    d. Temperament
    e. Adage

25. PRESAGE
    a. Tactful
    b. Viability
    c. Vow
    d. Prediction
    e. Exploit

26. ONEROUS
    a. Dubious
    b. Cultivate
    c. Arduous
    d. Squalid
    e. Emphatic

27. MUNIFICENT
    a. Candid
    b. Generous
    c. Livid
    d. Suitable
    e. Malleable

28. EXTANT
    a. Capitalize
    b. Surviving
    c. Foment
    d. Tentative
    e. Yield

29. DEFUNCT
    a. Simple
    b. Potent
    c. Geriatric
    d. Expend
    e. Extinct

30. CUPIDITY
    a. Happiness
    b. Greed
    c. Love
    d. Grief
    e. Anger

Answer Explanations

Verbal Analogies

1. C: This is a category analogy. Remember that we have to figure out the relationship between the first two words so that we can determine the relationship of the answer. Begonia is related to flower by *type*. Begonia is a type of flower, just as cardiologist is a type of doctor.

2. E: This is a synonym analogy. Notice that the word *malleable* is synonymous to the word *pliable*. Thus, in our answer, we should look for two words that have the same meaning. *Disparage* and *criticize* in Choice *E* have the same meaning, so this is the correct answer.

3. B: This is a part to whole analogy. The relationship between *cerebellum* and *brain* is that the cerebellum makes up part of the brain, while a *nucleus* makes up part of a *cell*.

4. A: This is an object to function analogy. Usually, a *whisk* is a cooking utensil used in the process of *baking*. As such, a *glove* is used in the sport of *boxing*. Both are objects used within a particular process.

5. C: This analogy relies on the logic of performer to related action. The original analogy says *umpires* (performer) *officiate* (action), which means to act as an official in a sporting event. In the same manner, a *counselor* (performer) *guides* (action) their clients toward well-being.

6. D: The analogy used here is degree/intensity. A *chuckle* is a giggle, while a *guffaw* is a burst of laughter. One is more intense than the other. *Whisper* is to talk softly, while *bellow* is to talk loudly. One is more intense than the other.

7. D: This analogy denotes a symbol and its representation. *Fire* can be representative of *passion*, while *ice* represents someone who is cold or *rigid*.

8. A: This is considered an antonym analogy. *Geriatric* means old age, and *youth* is the opposite of old age. Likewise, *transparent* means to see through something, while *opaque* means cloudy or muddy.

9. E: This is a cause and effect analogy. *Lying* causes *distrust,* while *hurricanes* cause *devastation*.

10. B: This is a synonym analogy. *Desolate* and *barren* both mean deserted. Likewise, *contend* and *maintain* are synonyms.

11. E: This is a part to whole analogy. Many *chapters* make up a *novel*, in the same way that many *strokes* make up a *painting*.

12. C: This is a performer to related action analogy. We know that *carpenters* perform *construction*, just as *wizards* perform *magic*.

13. C: This analogy relies on degree of intensity. *Tepid* means lukewarm, while *boiling* means extremely hot. In the same way, *rapacious* is an extreme form of *greed*.

14. C: This is a cause and effect analogy. To *read* is the cause or action, and a direct result that comes from reading is to *learn*. Likewise, when one *exercises* (cause), a direct result becomes better *health*.

15. A: This is an antonym analogy. *Competent* means capable of something, while *impotent* means incapable of something. In the same way, *demur* means to object, which is the opposite of *accept*.

16. B: This is a category/type analogy. *Merengue* is a type of *music,* just as *tension* is a type of *headache*.

17. D: This is an object/function analogy. The *car* (object) has the function of *transporting* people from one place to the other. Likewise, the function of a *fireplace* is to *heat* up a room.

18. E: This is a symbol/representation analogy. In artistic images or literature, *flowers* usually represent *femininity*. Likewise, traditionally in art and religion, the image or presence of *light* represents *transcendence*.

19. A: This is a synonym analogy. *Principle* means assumption or *truth,* while *shabby* means *squalid*.

20. E: This is a degree of intensity analogy. *Fastidious* means to be very *careful*, while *lament* means to *grieve* deeply. The second word in each of these is more intense than the first.

21. C: This is a part to whole analogy. *Mantle* is part of the four layers that make up *earth.* Likewise, a *bun* is one of the things that makes up a *hamburger.*

22. B: This is a performer to related action analogy. A *maestro* is one who is an expert musician and who *conducts* a musical performance. Likewise, an *acupuncturist* deals with natural *healing* within the body.

23. E: This is an antonym analogy. *Contentious* means not *agreeable,* while *pernicious* means stingy, or ungenerous.

24. C: This is a category/type relationship. *Cumulus* is a type of *cloud,* just as *tortellini* is a type of *pasta*.

25. A: This is an object/function analogy. One uses a *pot* to *boil.* Likewise, one uses a *belay* to *climb.*

26. B: This is a symbol/representation analogy. In Greek mythology, the *laurel* wreath is a symbol of *victory*. Additionally, in Christianity, a *dove* is used as a representation of *peace*.

27. D: This is a synonym analogy. *Transient* and *ephemeral* both mean temporary or short-lived. *Sanguine* means to be *cheerful.*

28. D: This is an antonym analogy. *Diverge* means to be at odds with, or the opposite of *agree.* Likewise, *antipathy* means a strong dislike or disgust, which is the opposite of *friendliness*.

29. E: This is a category/type analogy. *Nitrogen* is a specific type of *element*, just as *calico* is a type of *cat*.

30. C: This is a degree of intensity analogy. *Miniscule* is an extreme version of *small*, while *ailing* is an extreme version of *unhealthy*.

Synonyms

1. B: The word *garish* means excessively ornate or elaborate, which is most closely related to the word *flashy*. The word *drab* is the opposite of *garish*. The word *gait* means a particular manner of walking, *hardy* means robust or sturdy, and *lithe* means graceful and supple.

2. D: The word *inane* means *senseless* or absurd. The word *ratify* means to approve, *illicit* means illegal, *uncouth* means crude, and *wry* means clever.

3. E: The word *solace* most closely resembles the word *comfort*. *Marred* means scarred, *induce* means to cause something, *depose* means to dethrone, and *inherent* means natural.

4. D: *Copious* is synonymous with the word *ample*. *Dire* means urgent or dreadful, *adept* means skillful, *indignant* means angered by injustice, and *nuance* means subtle difference.

5. C: *Supercilious* means *arrogant*. *Tenuous* means weak or thin, *waning* means decreasing, *placate* means to appease, and *extol* means to praise or celebrate.

6. A: *Lurid* most closely resembles the word *gruesome*. *Placid* means calm or mild, *irate* means angry, *quell* means defeat or suppress, and *torpor* means lethargy.

7. B: *Vanquish* means to *conquer*. *Saturate* means to soak, *reproach* means to scold, *parch* means to make dry, and *surrender* is the opposite of *vanquish*.

8. B: *Trite* is closest to the word *banal*, which means common. *Scanty* means barely sufficient, *polemical* means controversial, *indulgent* means lenient, and *eclectic* means from diverse sources.

9. E: *Divulge* means to *admit* or confess something. *Dupe* means to deceive, *flummox* means to confuse, *indulgent* means lenient, and *germinate* means to grow.

10. D: *Indolent* and *lackadaisical* both mean lazy or indifferent. *Adamant* means unyielding, *dour* means gloomy or grim, *noisome* means bad or offensive, and *remiss* means careless or thoughtless.

11. C: *Bolster* and *support* are both synonyms. *Bequeath* means to hand down through a will, *abate* means to lessen, *palliate* means to remove pain, and *tractable* means easily managed.

12. B: *Unwitting* and *unintentional* are synonyms. *Undermine* means weaken, *rife* means excessively abundant, *pernicious* means harmful, and *stolid* means apathetic or impassive.

13. A: *Ungainly* means awkward or *clumsy*. *Absurd* means ridiculous or senseless, *unruly* means boisterous, *tenuous* means flimsy, and *petulant* means irritable.

14. A: *Prattle* and *babble* are synonyms; they both mean to talk incessantly. *Prosaic* means lacking imagination, *deluded* means tricked or betrayed, *meddle* means to intervene, and *folly* means silliness.

15. D: *Prolific* is most closely related to the word *productive*. *Devoid* means lacking, *elusive* means difficult to define, *laconic* means concise, and *judicious* means fair.

16. E: *Fortitude* means *courage*. *Aura* means air or character, *disparage* means to criticize or belittle, *finesses* means tact or know-how, and *cowardice* is the opposite of fortitude.

17. B: *Acumen* is most closely related to the word *ingenuity*. *Diligent* means hardworking, *congenial* means pleasant, *embroiled* means to be involved in an argument, and *reverent* means respectful or pious.

18. C: *Relegate* means *demote* or transfer. *Relay* means to pass on or transmit. *Temporize* means delay or evade, *vigilant* means watchful and alert, and *spurn* means to reject with contempt.

19. B: *Curtail* means to cut short or *abridge*. *Covet* means to crave something someone else has, *foil* means to hinder or prevent, *construe* means to make sense of something, and *bilk* means to cheat someone.

20. D: *Conserve* is most closely related to the word *maintain. Constrain* means to force or restrain, *adjourn* means to stop a proceeding, *stipulate* means to designate, and *improve* means to become better.

21. D: *Ambivalent* means to be *conflicted* or doubtful. *Accosted* means to approach for solicitation, *engrossed* means preoccupied with something, *impartial* means fair or unprejudiced, and *convinced* is the opposite of *ambivalent*.

22. C: *Disrepute* is most closely related to the word *dishonor. Benevolent* means kind or generous. *Condone* means to overlook or approve. *Emit* means to discharge. *Postulate* means to assert.

23. A: *Instigate* is most closely related to *provoke. Renounce* means to give up. *Prescribe* means to command orders. To *modify* means to change or alter. To *diminish* means to become smaller.

24. E: *Maxim* is a saying or *adage. Discord* is a disagreement. *Clout* means special advantage or power. *Apex* means the peak or highest point of a thing. *Temperament* means usual feelings or mood.

25. D: *Presage* is most closely related to the word *prediction. Tactful* is an adjective meaning skilled at dealing with people. *Viability* means done in a useful way. *Vow* is the same as a promise. *Exploit* means to use selfishly or for profit.

26. C: *Onerous* means difficult or *arduous. Dubious* means doubtful. *Cultivate* means to foster growth. *Squalid* means run-down or decrepit. *Emphatic* means using emphasis.

27. B: *Munificent* means *generous*. To be *candid* means to be honest or blunt. *Livid* means filled with rage. *Suitable* means to be fit for something. *Malleable* means pliable or adaptable.

28. B: The closest synonym to *extant* is *surviving. Extant* means to be in existence. *Capitalize* means using to one's own advantage. *Foment* means to stir up. *Tentative* means not finalized yet. *Yield* means to surrender.

29. E: *Defunct* most nearly means *extinct. Simple* means not complex. *Potent* means to have a lot of influence. *Geriatric* is relating to old age, so this isn't the best answer. To *expend* means to use up.

30. B: *Cupidity* means *greed* or strong desire, so the rest of the words do not fit as synonyms here.

Writing Sample

Parts of the Essay

The **introduction** has to do a few important things:

- Establish the **topic** of the essay in original wording (i.e., not just repeating the prompt)
- Clarify the significance/importance of the topic or purpose for writing. This should provide a brief overview rather than share too many details, a brief overview.
- Offer a **thesis statement** that identifies the writer's own viewpoint on the topic. Typically, the thesis statement is one or two brief sentences that offer a clear, concise explanation of the main point on the topic.

Body paragraphs reflect the ideas developed in the outline. Three or four points is probably sufficient for a short essay, and they should include the following:

- A **topic sentence** that identifies the sub-point (e.g., a reason why, a way how, a cause or effect)
- A detailed **explanation** of each sub-point, explaining why the writer thinks this point is valid
- **Illustrative examples**, such as personal examples or real-world examples, that support and validate the point (i.e., "prove" the point)
- A **concluding sentence** that connects the examples, reasoning, and analysis to the point being made

The **conclusion,** or final paragraph, should be brief and should reiterate the focus, clarifying why the discussion is significant or important. It is important to avoid adding specific details or new ideas to this paragraph. The purpose of the conclusion is to sum up what has been said to bring the discussion to a close.

The Short Overview

The essay may seem challenging, but following these steps can help writers focus:

- Take one-two minutes to think about the topic.
- Generate some ideas through brainstorming (three-four minutes).
- Organize ideas into a brief outline, selecting just three-four main points to cover in the essay
- Develop essay in parts:
    - Introduction paragraph, with intro to topic and main points
    - Viewpoint on the subject at the end of the introduction
    - Body paragraphs, based on outline, each should make a main point, explain the viewpoint, and use examples to support the point
    - Brief conclusion highlighting the main points and closing

- Read over the essay (last five minutes).
- Look for any obvious errors, making sure that the writing makes sense.

Writing an essay can be overwhelming, and performance panic is a natural response. The outline serves as a basis for the writing and helps to keep writers focused. Getting stuck can also happen, and it's helpful to remember that brainstorming can be done at any time during the writing process. Following the steps of the writing process is the best defense against writer's block.

Timed essays can be particularly stressful, but assessors are trained to recognize the necessary planning and thinking for these timed efforts. Using the plan above and sticking to it helps with time management. Timing each part of the process helps writers stay on track. Sometimes writers try to cover too much in their essays. If time seems to be running out, this is an opportunity to determine whether all of the ideas in the outline are necessary. Three body paragraphs are sufficient, and more than that is probably too much to cover in a short essay.

More isn't always *better* in writing. A strong essay will be clear and concise. It will avoid unnecessary or repetitive details. It is better to have a concise, five-paragraph essay that makes a clear point, than a ten-paragraph essay that doesn't. The goal is to write one-two pages of quality writing. Paragraphs should also reflect balance; if the introduction goes to the bottom of the first page, the writing may be going off-track or be repetitive. It's best to fall into the one-two page range, but a complete, well-developed essay is the ultimate goal.

Applying Basic Knowledge of the Elements of the Writing Process

Practice Makes Prepared Writers

Like any other useful skill, writing only improves with practice. While writing may come more easily to some than others, it is still a skill to be honed and improved. Regardless of a person's natural abilities, there is always room for growth in writing. Practicing the basic skills of writing can aid in preparations for the SSAT.

One way to build vocabulary and enhance exposure to the written word is through reading. This can be through reading books, but reading of any materials such as newspapers, magazines, and even social media count towards practice with the written word. This also helps to enhance critical reading and thinking skills, through analysis of the ideas and concepts read. Think of each new reading experience as a chance to sharpen these skills.

Planning

Brainstorming

One of the most important steps in writing an essay is prewriting. Before drafting an essay, it's helpful to think about the topic for a moment or two, in order to gain a more solid understanding of what the task is. Then, spending about five minutes jotting down the immediate ideas that could work for the essay is recommended. Brainstorming is a way to get some words on the page and offer a reference for ideas when drafting. Scratch paper is provided for writers to use any prewriting techniques such as webbing, free writing, or listing. The goal is to get ideas out of the mind and onto the page.

In the planning stage, it's important to consider all aspects of the topic, including different viewpoints on the subject. There are more than two ways to look at a topic, and a strong argument considers those opposing viewpoints. Considering opposing viewpoints can help writers present a fair, balanced, and informed essay that shows consideration for all readers. This approach can also strengthen an argument by recognizing and potentially refuting the opposing viewpoint(s).

Drawing from personal experience may help to support ideas. For example, if the goal for writing is a personal narrative, then the story should be from the writer's own life. Many writers find it helpful to draw from personal experience, even in an essay that is not strictly narrative. Personal anecdotes or short stories can help to illustrate a point in other types of essays as well.

Once the ideas are on the page, it's time to turn them into a solid plan for the essay. The best ideas from the brainstorming results can then be developed into a more formal outline.

Outlining

An **outline** is a system used to organize writing. When reading texts, outlining is important because it helps readers organize important information in a logical pattern using Roman numerals. Usually, outlines start out with the main idea(s) and then branch out into subgroups or subsidiary thoughts or subjects. The outline should be methodical, with at least two main points follow each by at least two subpoints. Not only do outlines provide a visual tool for readers to reflect on how events, characters, settings, or other key parts of the text or passage relate to one another, but they can also lead readers to a stronger conclusion. The sample below demonstrates what a general outline looks like.

I. Main Topic 1
    a. Subtopic 1
    b. Subtopic 2
        1. Detail 1
        2. Detail 2
II. Main Topic 2
    a. Subtopic 1
    b. Subtopic 2
        1. Detail 1
        2. Detail 2

Free Writing

Like brainstorming, **free writing** is another prewriting activity to help the writer generate ideas. This method involves setting a timer for 2 or 3 minutes and writing down all ideas that come to mind about the topic using complete sentences. Once time is up, review the sentences to see what observations have been made and how these ideas might translate into a more coherent direction for the topic. Even if sentences lack sense as a whole, this is an excellent way to get ideas onto the page in the very beginning stages of writing. Using complete sentences can make this a bit more challenging than brainstorming, but overall it is a worthwhile exercise, as it may force the writer to come up with more complete thoughts about the topic.

Writing

Now it comes time to actually write your essay. Follow the outline you developed in the brainstorming process and try to incorporate the sentences you wrote in the free writing exercise.

Basing the essay on the outline aids in both organization and coherence. The goal is to ensure that there is enough time to develop each sub-point in the essay, roughly spending an equal amount of time on each idea. Keeping an eye on the time will help. If there are fifteen minutes left to draft the essay, then it makes sense to spend about 5 minutes on each of the ideas. Staying on task is critical to success, and timing out the parts of the essay can help writers avoid feeling overwhelmed.

Remember that your work here does not have to be perfect. This process is often referred to as **drafting** because you're just creating a rough draft of your work.

Don't get bogged down on the small details. For instance, if you're not sure whether or not a word should be capitalized, mark it somehow and look up the capitalization rule while in the revision process if not in a testing situation. The same goes for referencing sources. That should not be focused on until after the writing process.

Referencing Sources

Anytime you quote or paraphrase another piece of writing you will need to include a citation. A **citation** is a short description of the work that your quote or information came from. The manual of style your teacher wants you to follow will dictate exactly how to format that citation. For example, this is how you would cite a book according to the APA manual of style:

- *Format*: Last name, First initial, Middle initial. (Year Published) *Book Title.* City, State: Publisher.
- *Example*: Sampson, M. R. (1989). *Diaries from an Alien Invasion. Springfield, IL*: Campbell Press.

Revising

Revising and proofreading offers an opportunity for writers to polish things up. Putting one's self in the reader's shoes and focusing on what the essay actually says helps writers identify problems—it's a movement from the mindset of writer to the mindset of editor. The goal is to have a clean, clear copy of the essay.

During the essay portion of a test, leaving a few minutes at the end to revise and proofread offers an opportunity for writers to polish things up. Putting one's self in the reader's shoes and focusing on what the essay actually says helps writers identify problems—it's a movement from the mindset of writer to the mindset of editor. The goal is to have a clean, clear copy of the essay. The following areas should be considered when proofreading:

- Sentence fragments
- Awkward sentence structure
- Run-on sentences
- Incorrect word choice
- Grammatical agreement errors
- Spelling errors
- Punctuation errors
- Capitalization errors

Recursive Writing Process

While the writing process may have specific steps, the good news is that the process is recursive, meaning the steps need not be completed in a particular order. Many writers find that they complete steps at the same time such as drafting and revising, where the writing and rearranging of ideas occur simultaneously or in very close order. Similarly, a writer may find that a particular section of a draft needs more development, and will go back to the prewriting stage to generate new ideas. The steps can be repeated at any time, and the more these steps of the recursive writing process are employed, the better the final product will be.

Developing a Well-Organized Paragraph

Forming Paragraphs

A good **paragraph** should have the following characteristics:

- Be logical with organized sentences
- Have a unified purpose within itself
- Use sentences as building blocks
- Be a distinct section of a piece of writing
- Present a single theme introduced by a topic sentence
- Maintain a consistent flow through subsequent, relevant, well-placed sentences
- Tell a story of its own or have its own purpose, yet connect with what is written before and after
- Enlighten, entertain, and/or inform

Though certainly not set in stone, the length should be a consideration for the reader's sake, not merely for the sake of the topic. When paragraphs are especially short, the reader might experience an irregular, uneven effect; when they're much longer than 250 words, the reader's attention span, and probably their retention, is challenged. While a paragraph can technically be a sentence long, a good rule of thumb is for paragraphs to be at least three sentences long and no more than ten sentence long. An optimal word length is 100 to 250 words.

Coherent Paragraphs

Coherence is simply defined as the quality of being logical and consistent. In order to have coherent paragraphs, therefore, authors must be logical and consistent in their writing, whatever the document might be. Two words are helpful to understanding coherence: flow and relationship. Earlier, transitions were referred to as being the "glue" to put organized thoughts together. Now, let's look at the topic sentence from which flow and relationship originate.

The **topic sentence**, usually the first in a paragraph, holds the essential features that will be brought forth in the paragraph. It is also here that authors either grab or lose readers. It may be the only writing that a reader encounters from that writer, so it is a good idea to summarize and represent ideas accurately.

The coherent paragraph has a logical order. It utilizes transitional words and phrases, parallel sentence structure, clear pronoun references, and reasonable repetition of key words and phrases. Use common sense for repetition. Consider synonyms for variety. Be consistent in verb tense whenever possible.

When writers have accomplished their paragraph's purpose, they prepare it to receive the next paragraph. While writing, read the paragraph over, edit, examine, evaluate, and make changes accordingly. Possibly, a paragraph has gone on too long. If that occurs, it needs to be broken up into other paragraphs, or the length should be reduced. If a paragraph didn't fully accomplish its purpose, consider revising it.

Main Point of a Paragraph

What is the main point of a paragraph? It is *the* point all of the other important and lesser important points should lead up to, and it should be summed up in the topic sentence.

Sometimes there is a fine line between a paragraph's topic sentence and its main point. In fact, they actually might be one and the same. Often, though, they are two separate but closely related aspects of the same paragraph.

Depending upon writer's purpose, they might not fully reveal the topic sentence or the paragraph's main point until the paragraph's conclusion.

Sometimes, while developing paragraphs, authors deviate from the main point, which means they have to delete and rework their materials to stay on point.

Examining Paragraphs

Throughout this text, composing and combining sentences, using basic grammar skills, employing rules and guidelines, identifying differing points of view, using appropriate context, constructing accurate word usage, and discerning correct punctuation have all been discussed. Whew! The types of sentences, patterns, transitions, and overall structure have been covered as well.

While authors write, thoughts coalesce to form words on "paper" (aka a computer screen). Authors strategically place those thoughts in sentences to give them "voice" in an orderly manner, and then they manipulate them into cohesive sentences for cohesion to express ideas. Like a hunk of modeling clay (thanks to computers, people are no longer bound to erasers and whiteout), sentences can be worked and reworked until they cooperate and say what was originally intended.

Before calling a paragraph complete, identify its main point, making sure that related sentences stay on point. Pose questions such as, "Did I sufficiently develop the main point? Did I say it succinctly enough? Did I give it time to develop? *Is* it developed?"

Let's examine the following two paragraphs, each an example of a movie review. Read them and form a critique.

> Example 1: *Eddie the Eagle* is a movie about a struggling athlete. Eddie was crippled at birth. He had a lot of therapy and he had a dream. Eddie trained himself for the Olympics. He went far away to learn how to ski jump. It was hard for him, but he persevered. He got a coach and kept trying. He qualified for the Olympics. He was the only one from Britain who could jump. When he succeeded, they named him, "Eddie the Eagle."

> Example 2: The last movie I saw in the theater was *Eddie the Eagle,* a story of extraordinary perseverance inspired by real life events. Eddie was born in England with a birth defect that he slowly but surely overcame, but not without trial and error (not the least of which was his father's perpetual *dis*couragement). In fact, the old man did everything to get him to give up, but Eddie was dogged beyond anyone in the neighborhood; in fact, maybe beyond anyone in the whole town or even the whole world! Eddie, simply, did not know to quit. As he grew up, so did his dream; a strange one, indeed, for someone so unaccomplished: to compete in the Winter Olympics as a ski jumper (which he knew absolutely nothing about). Eddie didn't just

keep on dreaming about it. He actually went to Germany and *worked* at it, facing unbelievable odds, defeats, and put-downs by Dad and the other Men in Charge, aka the Olympic decision-makers. Did that stop him? No way! Eddie got a coach and persevered. Then, when he failed, he persevered some more, again and again. You should be able to open up a dictionary, look at the word "persevere," and see a picture of Eddie the Eagle because, when everybody told him he couldn't, he did. The result? He is forever dubbed, "Eddie the Eagle."

Both reviews tell something about the movie *Eddie the Eagle*. Does one motivate the reader to want to see the movie more than the other? Does one just provide a few facts while the other paints a virtual picture of the movie? Does one give a carrot and the other a rib eye steak, mashed potatoes, and chocolate silk pie?

Paragraphs sometimes only give facts. Sometimes that's appropriate and all that is needed. Sometimes, though, writers want to use the blank documents on their computer screens to paint a picture. Writers must "see" the painting come to life. To do so, pick a familiar topic, write a simple sentence, and add to it. Pretend, for instance, there's a lovely view. What does one see? Is it a lake? Try again—picture it as though it were the sea! Visualize a big ship sailing out there. Is it sailing away or approaching ? Who is on it? Is it dangerous? Is it night and are there crazy pirates on board? Uh-oh! Did one just jump ship and start swimming toward shore?

Distinguishing Between Formal and Informal Language

It can be helpful to distinguish whether a writer or speaker is using formal or informal language because it can give the reader or listener clues to whether the text is informative, nonfiction, argumentative, or the intended tone or audience. Formal and informal language in written or verbal communication serve different purposes and are often intended for different audiences. Consequently, their tone, word choices, and grammatical structures vary. These differences can be used to identify which form of language is used in a given piece and to determine which type of language should be used for a certain context. Understanding the differences between formal and informal language will also allow a writer or speaker to implement the most appropriate and effective style for a given situation.

Formal language is less personal and more informative and pragmatic than informal language. It is more "buttoned-up" and business-like, adhering to proper grammatical rules. It is used in professional or academic contexts, to convey respect or authority. For example, one would use formal language to write an informative or argumentative essay for school and to address a superior or esteemed professional like a potential employer, a professor, or a manager. Formal language avoids contractions, slang, colloquialisms, and first person pronouns. **Slang** refers to non-standard expressions that are not used in elevated speech and writing. Slang creates linguistic in-groups and out-groups of people, those who can understand the slang terms and those who can't. Slang is often tied to a specific time period. For example, "groovy" and "far out" are connected to the 1970s, and "as if!" and "4-1-1-" are connected to the 1990s. **Colloquial language** is language that is used conversationally or familiarly—e.g., "What's up?"—in contrast to formal, professional, or academic language—"How are you this evening?" Formal language uses sentences that are usually more complex and often in passive voice. Punctuation can differ as well. For example, **exclamations point (!)** are used to show strong emotion or can be used as an interjection but should be used sparingly in formal writing situations.

Informal language is often used when communicating with family members, friends, peers, and those known more personally. It is more casual, spontaneous, and forgiving in its conformity to grammatical rules and conventions. Informal language is used for personal emails, some light fiction stories, and some correspondence between coworkers or other familial relationships. The tone is more relaxed and slang, contractions, clichés, and the first and second person may be used in writing. The imperative voice may be used as well.

As a review, the perspectives from which something may be written or conveyed are detailed below:

- First-person point of view: The story is told from the writer's perspective. In fiction, this would mean that the main character is also the narrator. First-person point of view is easily recognized by the use of personal pronouns such as *I, me, we, us, our, my*, and *myself.*
- Second-person point of view: This point of view isn't commonly used in fiction or nonfiction writing because it directly addresses the reader using the pronouns *you, your*, and *yourself.* Second-person perspective is more appropriate in direct communication, such as business letters or emails.
- Third-person point of view: In a more formal essay, this would be an appropriate perspective because the focus should be on the subject matter, not the writer or the reader. Third-person point of view is recognized by the use of the pronouns *he, she, they*, and *it*. In fiction writing, third person point of view has a few variations.
    - Third-person limited point of view refers to a story told by a narrator who has access to the thoughts and feelings of just one character.
    - In third-person omniscient point of view, the narrator has access to the thoughts and feelings of all the characters.
    - In third-person objective point of view, the narrator is like a fly on the wall and can see and hear what the characters do and say but does not have access to their thoughts and feelings.

Writing Prompts

In the SSAT Upper Level Writing Sample, you will have a choice between two prompts. One prompt will be a traditional essay, and one will be a creative essay. The organization of your ideas is very important for both these essays. The essay should be clear and coherent regardless of which prompt you choose. Two sample prompts are available below.

Argument Prompt

Write an argumentative essay. Take a side on the following question:

> Currently, what is the biggest issue we face as human beings and how should we go about solving it?

Creative Prompt

Write a creative essay that begins with the phrase below.

> The air inside had suddenly become chilly . . .

Dear SSAT Upper Test Taker,

We would like to start by thanking you for purchasing this study guide for your SSAT Upper exam. We hope that we exceeded your expectations.

Our goal in creating this study guide was to cover all of the topics that you will see on the test. We also strove to make our practice questions as similar as possible to what you will encounter on test day. With that being said, if you found something that you feel was not up to your standards, please send us an email and let us know.

We would also like to let you know about other books in our catalog that may interest you.

PSAT 8/9

This can be found on Amazon: amazon.com/dp/162845976X

SAT

amazon.com/dp/1628458984

ACT

amazon.com/dp/1628458844

ACCUPLACER

amazon.com/dp/162845945X

CLEP College Composition

amazon.com/dp/1628454199

We have study guides in a wide variety of fields. If the one you are looking for isn't listed above, then try searching for it on Amazon or send us an email.

Thanks Again and Happy Testing!
Product Development Team
info@studyguideteam.com

FREE Test Taking Tips DVD Offer

To help us better serve you, we have developed a Test Taking Tips DVD that we would like to give you for FREE. **This DVD covers world-class test taking tips that you can use to be even more successful when you are taking your test.**

All that we ask is that you email us your feedback about your study guide. Please let us know what you thought about it – whether that is good, bad or indifferent.

To get your **FREE Test Taking Tips DVD**, email freedvd@studyguideteam.com with "FREE DVD" in the subject line and the following information in the body of the email:

a. The title of your study guide.

b. Your product rating on a scale of 1-5, with 5 being the highest rating.

c. Your feedback about the study guide. What did you think of it?

d. Your full name and shipping address to send your free DVD.

If you have any questions or concerns, please don't hesitate to contact us at freedvd@studyguideteam.com.

Thanks again!

CPSIA information can be obtained
at www.ICGtesting.com
Printed in the USA
LVHW100848150921
697803LV00018B/321

9 781628 457827